Testing

TESTING

Theoretical and Applied Perspectives

Edited by
RONNA F. DILLON and
JAMES W. PELLEGRINO

PRAEGER

New York
Westport, Connecticut
London

The views and conclusions contained in this document are those of the authors and should not be interpreted as necessarily representing the official policies of the sponsoring agencies or the U.S. government.

Library of Congress Cataloging-in-Publication Data

Testing : theoretical and applied perspectives / edited by Ronna F.
    Dillon and James W. Pellegrino.
        p. cm.
    Bibliography: p.
    Includes index.
    ISBN 0-275-92759-8 (alk. paper)
    1. United States—Armed Forces—Examinations. 2. Military
research—United States. 3. Educational tests and measurements.
4. Intelligence tests. 5. Psychological tests. I. Dillon, Ronna
F. II. Pellegrino, James W.
UB336.T47   1988
355'0076—dc 19   88-316

Library of Congress Catalog Card Number: 88-316
ISBN: 0-275-92759-8

First published in 1989

Praeger Publishers, One Madison Avenue, New York, NY 10010
A division of Greenwood Press, Inc.

Printed in the United States of America

The paper used in this book complies with the
Permanent Paper Standard issued by the National
Information Standards Organization (Z39.48-1984).

10 9 8 7 6 5 4 3 2 1

Contents

Tables and Figures

TABLES

FIGURES

Preface

Research aimed at developing new approaches to testing and training of complex intellectual occupies the forefront of psychology and education. Theoretical and applied concerns provide motivation for research in both areas. This book is intended to highlight some of the ongoing efforts being pursued by the military research and development community.

With respect to research on testing, accurate assessment of cognitive abilities and processes is essential if we are to gain information about subtheories and theories of intellectual abilities. Powerful testing methods and measures also are essential if we are to predict school and job performance efficiently and effectively in civilian and military sectors.

Advances in testing covered in this book include the use of item calibration procedures and new administration techniques for use with existing ability batteries such as the Armed Forces Vocational Aptitude Battery (ASVAB). Also included is the development of computerized tests that can administer and score memory items, dynamic or moving tasks, and timed items much more effectively than can be accomplished using paper and pencil tests. Testing approaches that make use of new recording methods, such as psychophysiological and electrophysiological techniques, to derive measures of information processing during task, solution are also described.

Preparation of this book was a major collaborative effort supported by the Navy Personnel Research and Development Center in San Diego, California. Researchers at major military personnel research organizations and universities were invited to contribute summaries of their work. Research conducted at the Air Force Human Rsources Laboratory, the Army Research Institute for the Behavioral and Social Sciences, and the Navy Personnel Research and Development Center is among work summarized in this book.

Several individuals played major roles in coordinating this effort. Dr. John J. Pass, of the Navy Personnel Research and Development Center, was the primary initiator and coordinator for this publication. Drs. Edwin G. Aiken, Richard C. Sorenson, and Pat-Anthony Federico, also from the Navy Personnel Research and Development Center, were highly facilitative in producing this publication.

Preparation of this document was funded by the Office of the Assistant Secretary of Defense (Accessions Directorate), the Air Force Office of Scientific Research, a faculty fellowship from the American Society for Engineering Education, and a Spencer Fellowship from the National Academy of Education.

Testing

1 New Approaches to Aptitude Testing

Ronna F. Dillon

Research in testing is centered on elucidating intellectual processes and representations as well as enhancing prediction of the outcomes of both education/training and job performance. Accurate or sensitive measurement of mental abilities forms the basis of attempts to verify subtheories or theories of intelligence. Similarly, accurate measurement is essential if psychologists, educators, and employers are to predict satisfactorily who will benefit most from certain types of training as well as who will perform maximally on the job.

Traditional psychometric approaches fall short in two respects. These techniques do not account for satisfactory amounts of variance in criterion measures of interest (for example Dillon, 1985a, 1986; Jencks, 1977; Matarazzo, 1972;

Preparation of this chapter was supported by contract TCN 86-144 provided by the Navy Personnel Research and Development Center, through a Scientific Services Agreement issued by Battelle Columbus Labatories.

1

McCall, 1977; Wechsler, 1975). Moreover, traditional methods are diagnostically and prescriptively barren (Dillon, 1986; Hunt, 1983; Sternberg, 1981).

This chapter is divided into two sections. The first section contains a discussion of new approaches to aptitude or ability testing. In the second section, I provide a set of criteria against which I evaluate the various testing paradigms.

NEW TESTING APPROACHES

A host of new testing approaches have been conceptualized. These approaches involve several innovations. One set of testing approaches involves use of new methods of test administration. A second set of approaches involves altering the measures taken from the test, while a third set of paradigms centers on conceptualizing new aptitude dimensions. Within this third category, some researchers have studied familiar aptitudes that have not been used for predictive purposes; other investigators have developed new measures of familiar aptitudes; other researchers have varied the theoretical organization among dimensions; while other researchers have conceptualized extensions of traditional dimensions (for example, postulating the significance of *dynamic* in addition to static spatial ability); and still others have conceptualized new dimensions. The expanded predictor space is depicted in Figure 1.1.

Altering the Method of Test Administration

With respect to altering the method of test administration, the value of dynamic testing approaches has been acknowledged. Two different definitions of intelligence underlie two correspondingly different testing purposes. One type of activity is motivated by a view that intelligence is a person's current level of cognitive competence. According to this view, intelligence is what a person really *knows*, as distinct from *performance*, which is some percentage of a person's competence that is manifested behaviorally. In this paradigm, the goal of testing is to bring performance in line with competence. Testing techniques often involve mental elaborations that involve augmented feedback mechanisms and/or dual-coding during information processing. The approach involves examinees in a testing situation that is similar to the manner in which thinking and learning occur. This conception of intelligence, and the testing techniques spawned therefrom, is of particular interest to scientists or diagnosticians whose objective it is to ascertain whether an individual possesses the requisite knowledge to perform satisfactorily in a given educational program or, conversely, whether an individual's competence is such that remedial activities are warranted.

Research involving this type of dynamic testing has yielded a number of interesting findings. First, both performance and criterion-related validity increase when verbalization or elaborated feedback is used during administration of complex reasoning items (for example, Carlson & Dillon,

Figure 1.1
Aptitude Predictor Space

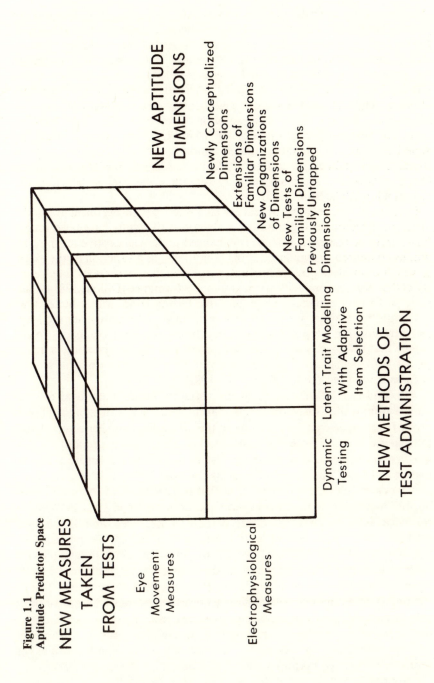

NEW MEASURES
TAKEN
FROM TESTS

Eye
Movement
Measures

Electrophysiological
Measures

NEW APTITUDE
DIMENSIONS

Newly Conceptualized
Dimensions

Extensions of
Familiar Dimensions

New Organizations
of Dimensions

New Tests of
Familiar Dimensions
Previously Untapped
Dimensions

Dynamic   Latent Trait Modeling
Testing   With Adaptive
             Item Selection

NEW METHODS OF
TEST ADMINISTRATION

1979a). Second, the extent to which elaborative procedures are necessary to activate cognitive competence (that is, what a person really knows) is a function of the experimental background of the examinee and the relationship of learner characteristics to task demands. With respect to experiential background, my colleagues and I (for example, Carlson & Dillon, 1979b; Dillon, 1979) reported that inductive reasoning performance of hearing-impaired children was enhanced significantly when examinees were instructed to "think aloud" during solution of test items. Similarly, superior inductive reasoning performance of white children vis-à-vis Mexican-American and black children under standard testing procedures was not found under elaborative testing procedures, wherein performance of the Mexican-American and black subjects went up to the level of the white subjects (Dillon & Carlson, 1978).

In terms of the relation of learner characteristics to performance under different dynamic testing conditions, two types of results have been found. First, aptitude test performance is enhanced under elaborative testing conditions compared to standard testing conditions, as a function of the extent to which examinees possess attributes (for example, aspects of intellectual style) that are not conducive to maximum performance (Dillon, 1981a). Second, performance under one elaborative testing condition is better than under another elaborative condition (for example, internally structured subject verbalization versus externally structured examiner feedback) to the degree that a subject's preferred mode of structuring information is inconsistent with demands of one or more of the particular elaborative procedures being used (Dillon & Donow, 1982).

A second type of activity involving variation in the method of test administration is motivated by a view that intelligence is a person's ability to profit from experiences. According to this perspective, rather than viewing intelligence as a person's "current" level of competence, proponents view intelligence as a latent competence that must be trained before it can be manifested fully. The notion underlying this approach is that certain people may lack test-relevant experiences, so that true intellectual competence will not be manifested until these experiential deficiencies are ameliorated. In this sense, intelligence is "changed" before it is measured, and this changed intellectual performance is believed to be a person's actual competence. The goal of this intelligence-training work is to equip a person with task-relevant skills that normally are acquired in the course of academic and life experiences and then to test the individual's intellectual competence. Testing techniques often involve several directed practice sessions with one or more of a variety of types of tasks, often similar to the actual test tasks. This second approach to intelligence testing may be useful in situations where the objective of the testing is to ascertain the likelihood that an individual will profit from (that is, master) a course of academic or occupational training. The approach may be useful, for example, when socioculturally different individuals are admitted to a technical training program. Some cognitive-training researchers (Embretson, in press; Feuerstein, 1979; Vygotsky, 1978) believe that intelligent information processing must be internalized through training before accurate assessment of intellectual competence is possible.

In a third set of procedures that involve varying the method of test administration, latent trait models and adaptive item selection are employed (Embretson, 1983, 1984, 1985, 1986a). The basic issue in this type of research is to test hypotheses about information processing from mathematical models of task response, time, and accuracy. In this research, mathematical models are used to operationalize theories about the cognitive components that underlie test item responses. Theories then are evaluated, in part, by the fit of models for the particular types of data from a given set of examinees. The work is designed to minimize test length while preserving validity. Embretson discusses the application of component latent trait models to existing reasoning and comprehension test items to permit connecting item responses to cognitive theories of abilities. As with other componential work, specification of examinee component abilities, through varying item content, is possible. The work's theoretical strength, like other cognitive components work, is that understanding the cognitive characteristics of items that measure particular abilities facilitates understanding the nature of those abilities.

This work is aimed toward clarifying the construct validity of a test by addressing the two separate research stages of construct representation and nomothetic span (Embretson, 1983). Construct representation centers on identification of the constructs—components, strategies, and knowledge structures—involved in performance on a particular set of test items. Nomothetic span concerns the usefulness of the test as a measure of individual differences. With respect to construct representation, component latent trait models are used by Embretson to perform a type of cognitive component analysis of aptitude. As with other componential work (Dillon, 1986), collection of experimental evidence (for example, reaction time analyses of components, analysis of eye movement patterns for isolated components) precedes calibration of models. Calibration of component latent trait models follows postulation of a theory of item solving. This theory includes specification of the stimulus factor impact on the components.

In evaluating nomothetic span, interest is in explicating how the components influence the relationship of the test score to other measures of individual differences (Embretson, 1983). Tests with different weightings of the components have a different pattern of nomothetic span and different predictive validity.

Altering the Measures Taken from Aptitude Tests

A second class of new aptitude testing approaches involves altering the measures taken from the test. Here, the same aptitude measures (or different measures) may be administered. Emphasis is not on the characteristics of the test, rather concern is on collecting measures of ongoing information processing, during solution of test items, in place of test scores. This work is based on the premise that a small number of elementary processes underlie performance across a broad range of tasks requiring intelligent behavior. One example of work in which the measures taken from the test are varied is work conducted in my laboratory (for example, Dillon, 1981b, 1985a, 1987a, 1987b; Dillon & Reznick, 1986). The research involves using measures of information processing, derived from collection of eye movement data during solution of intact test items, to enhance understanding of the nature of intelligence and to improve prediction of intelligent performance. With respect to the goal of increasing our understanding of the nature of intelligence, I have provided information that helps elucidate components (for example, Dillon, 1981b, 1987a) and metacomponents (Dillon, 1987b) of intelligence. Regarding the goal of improving prediction of intelligent behavior, prediction models based on information processing measures, derived from eye movement data, account for as much as 80 percent of the variance in academic and technical training program performance and job performance, compared to 9 to 42 percent of the variance accounted for by *scores* taken from the same tests from which the eye movement records are derived (Dillon, 1985b; Dillon & Reznick, 1986).

A second example of the general approach of using information processing techniques to understand and predict intelligent performance involves the use of electrophysiological indexes of processing as measures of intelligence. I have evaluated this work in detail elsewhere (Dillon, 1987c). Basically, the approach involves using neuroelectric (that is, electroencephalographic [EEG], or evoked potential [EP]) or neuromagnetic (evoked fields [EF]) recording to elucidate individual differences in information processing and to relate these differences to external criteria, such as measures of academic achievement or on-job performance.

The use of neuroscience procedures in an attempt to measure directly brain processes, for the purpose of understanding and predicting intelligent behavior, is still exploratory in nature. Nevertheless, the procedures have yielded interesting preliminary findings with respect to prediction. Data indicate that it may be possible to use such measures to enhance performance prediction for highly technical Navy jobs such as sonar operation, and among Navy pilots and radar intercept officers as well as among Marine Corps security guards (Lewis, 1983; Lewis et al., 1986). Moreover, EP data have shown promise for predicting language performance during early childhood (for example, Molfese & Molfese, 1985). This predictive power may result from the use of information processing measures that tap processes common to predictor and criterion measures.

Use of Measures of New Aptitude Dimensions

In addition to varying the method of test administration and varying the measures taken from the test, work currently is underway to conceptualize and use measures of new aptitudes in place of more narrowly defined aptitudes. Research in this area falls into one of five categories. These categories are (1) familiar but heretofore untapped aptitude dimensions; (2) familiar dimensions tapped by new kinds of tests; (3) new organizations of existing aptitude dimensions; (4) extensions of aptitude dimensions; and (5) newly conceptualized aptitude dimensions.

Previously Untapped Dimensions

One set of studies involves using familiar aptitudes that have not previously been included in batteries used to predict academic or occupational success. And example of this type of work is the large-scale effort underway, headed by Mitchell, to use writing skills as a component in medical school admission (Association of American Medical Colleges, 1985). Mitchell notes that skills assessed in this work include developing a central idea, synthesizing concepts and other ideas, separating relevant from irrelevant information, developing alternative hypotheses, presenting ideas in a cohesive and logical manner, and manifesting writing skills that demonstrate clear exposition that uses the accepted practices of grammar, syntax, punctuation, and spelling. Attention to audience and originality also are evaluated in some admission and achievement testing programs. This effort reflects an increasingly popular belief that the composition skills tapped therein are important to success in medical school.

Related work being conducted by staff working on the National Assessment of Education Progress (Mullis, 1985) reflects acknowledgment of the multidimensional nature of writing demands. Clearly, there are different purposes for writing and different life tasks of which writing is a part. For example, writing may be informational, persuasive, or imaginative.

Familiar Dimensions Tapped by New Kinds of Tests

A second type of work in the category of new aptitudes centers on the use of new types of tests for assessing a familiar dimension. For example, the National Board of Medical Examiners (NBME) assesses problem-solving ability for patient management by means of written clinical simulation exams. Similarly, simulated patient exams are being used with increasing frequency in all areas of medical education.

New Organizations of Dimensions

A third type of research involves new organizations of existing aptitude dimensions. This research is directed toward clarifying the nature and organization of mental abilities. Snow and Lohman (1984) consider the importance of such clarification for the educational process. The importance of particular sets of aptitudes to intelligent behavior is an issue of widespread theoretical and applied significance. The work of Kyllonen and Christal and other members of the Learning Abilities Measurement

Program (LAMP) is characteristic of this type of undertaking. Kyllonen (1986) has proposed a four-source framework for understanding learning skills. According to the framework, learning results from individual differences in (1) effective cognitive processing speed; (2) effective working memory capacity; (3) conceptual knowledge; and (4) procedural and strategic knowledge. With respect to processing speed, Kyllonen posits possible cognitive components underlying solution of the different types of cognitive speed tests and advocates an analysis of stage duration that relies on a part correlation method wherein the latency of a particular component is estimated by statistically holding constant the duration of all predecessor tasks. Models are verified by comparing stage latencies with performance on intact tasks. The LAMP group is attempting to collect data to support their hypothesis that there are different kinds of processing speed, operating in different thinking and learning contexts. With respect to working memory capacity, the LAMP group is collecting data on the effect of working memory capacity on attentional and learning capabilities and on more conventional aptitude tests as well as on the possible multidimensional nature of the construct. Data collecton is beginning on the other two factors in the LAMP battery.

Extensions of Aptitude Dimensions

A fourth type of new aptitudes work involves extensions of familiar aptitude dimensions. An example of this type of research is the work of Pellegrino and Hunt on dynamic spatial processing. Studies of spatial ability have been conducted for several years, and a number of information processing analyses have been reported (for example, Cooper & Mumaw, 1985; Pellegrino & Goldman, 1983). Pellegrino and Hunt's extension of spatial ability involves examining dynamic spatial reasoning. This work provides a nice example of extrapolations of constructs that not only broaden knowledge regarding the nature of the ability under study but also yield aptitude constructs that may more closely parallel the role of these abilities in intelligent performance in the everyday world. Work of this type demonstrates the manner in which advances in knowledge about human abilities must await relevant technological progress. Moreover, the work takes a componential approach to large-scale computer-based testing. Efforts to predict external performance thereby may be enhanced.

Newly Conceptualized Dimensions

A fifth type of work centering on new aptitude dimensions involves newly conceptualized dimensions. Work in this area includes assessments of practical intelligence (for example, Sternberg & Wagner, 1986) and the multidimensional construct of cognitive flexibility (Dillon, 1987a, 1987b). Work in these two types of dimensions seems to be particularly promising with respect to predicting academic and occupational success among high-performing individuals. My rationale here is that considerable range restriction occurs for these individuals on aptitudes tapping explicitly acquired knowledge and skills. If additional variance is to be targeted, reasonably it might come from aptitude areas where the skills must be acquired tacitly; that is, during the course of training or occupational duties. With respect to tacitly acquired occupational knowledge, my colleagues and I (Dillon, Wagner, Loschen, & Travis, 1987) have collected data on psychiatry clerks that indicates that general and specific tacitly acquired occupational knowledge is an important component of success in advanced medical training for these very high-performing individuals. Regarding cognitive flexibility, data from my lab indicate that the ability to generate multiple strategies in solution of inductive reasoning items that can be solved in more than one way is an important component of academic success among advanced undergraduates (Dillon, 1987d) and among psychiatry clerks (Dillon, Loschen, & Travis, 1987). Other newly conceptualized aptitude dimensions that appear particularly promising are abilities to deal with novelty and to automatize information processing (Sternberg, 1985) and Bovens and Bohrer's work on expanding the use of affective measures as predictors of performance.

EXPANDING CRITERION DEVELOPMENT

My comments thus far have centered on new testing approaches designed to be more effective by expanding or reconceptualizing the predictor space. When trying to understand intelligent behavior, it becomes important to look at robust manifestations of intelligent functioning. Such a goal underlies recent attempts to broaden the criterion space; that is, to use broader and more realistic criterion measures.

A number of exciting advances are occurring that focus on improving systems of criterion development. In the medical sphere, Williams (1987) reports on a large-scale effort to develop and implement a comprehensive performance-based clinical exam for assessing performance in medical school. This exam measures student application of knowledge and skill in solving standardized patient problems, and the test is designed to duplicate circumstances and constraints experienced in clinical settings. The need to broaden the criterion space is addressed in this work.

Other research on criterion development is designed to provide laboratory-based learning opportunities, such as the lab being established by members of the LAMP group. An advantage of this work over traditional means of acquiring criterion data is the ready availability of criterion data in close temporal proximity to the predictor data, without requiring assistance from personnel at training sites.

A third advance in criterion development centers on use of job sample data in place of paper and pencil end-of-course exams. While these measures constitute a tremendous leap in terms of developing valid criterion measures, potential problems could render their use difficult. First, collection of job sample data is an extremely expensive and time-consuming endeavor. Second, strides must be made toward setting valid competency standards and dealing with range restriction and toward reducing subjectivity and improving inter-judge reliability among raters. A framework for understanding criterion development is discussed by Kevin Murphy in chapter 10.

EVALUATION OF TESTING APPROACHES

New testing approaches can be evluated both with respect to each approach's theoretical motivation and in terms of the practical utility of the techniques. Issues relevant to both theoretical and practical significance are addressed below.

Theoretical Motivation

With respect to new approaches to aptitude testing, issues of interest include theoretical underpinnings such as the manner in which the method connects with theories of abilities, and the sorts of questions about human abilities that the method is capable of addressing.

The component latent trait approach used by Embretson is promising in this regard. Embretson (1983, 1984, 1985, 1986) derives componential models from combinations of stages of specific difficulties. Similarly, I have reported a number of studies (for example, Dillon, 1981b, 1985a, 1985b, 1986, 1987a, 1987b) in which data about information processing stages are derived from examination of eye movement patterns during solution of items wherein stage difficulties have been manipulated. This work provides direct linkages between the measures derived from the approach and underlying theory. Members of the LAMP group also are collecting data designed to provide information about components of cognitive tasks by comparing stage durations on tasks comprised of specific sets of componential operations. Data are being collected by LAMP investigators on processing speed, working memory, declarative knowledge, and procedural knowledge. Also, as noted with regard to dynamic spatial ability, computerized test administration permits investigation of aptitude dimensions heretofore not amenable to study.

Componential work, which occupies a significant place in the forefront of intelligence theory and testing, has the strength of tapping fundamental information processing activities. As such, the work addresses robust processes that underlie successful performance across a broad range of tasks. Understanding these processes is contributing significantly to improving researchers' efforts to understand the nature of intellectual abilities.

A related concern centers on the robustness of aptitude constructs. Focus here is on the relevance of constructs and the tests designed to measure them across academic disciplines and across occupational and other life domains.

Considerable interest exists with regard to expanding testing efforts to occupational, professional, and adaptive behavior contexts (for example, Sternberg & Wagner, 1986). Methods and measures that do equally well in predicting performance among low-performing college freshmen (Anderson & Dillon, 1987) and advanced medical students (Dillon, Loschen, & Travis, 1987; Dillon, Wagner, Loschen, & Travis, 1987) are needed.

In conceptualizing aptitude constructs, researchers also need to consider the role of experience in proficiency. For example, it is quite possible that certain individual difference dimensions are important predictors of occupational success during early stages of training, while other aptitudes become important when greater amounts of expertise (and, concomitantly, greater amounts of responsibility) are acquired. The work of Appel and Watson (1987) addresses this point with respect to career development among Air Force officers.

Practical Utility

Issues regarding the practical utility of new testing approaches include routine psychodiagnostic utility, ease of implementation, replicability/standardizability, and criterion-related validity. These issues are discussed below.

Psychodiagnostic Utility

The concern here is the extent to which meaningful comparisons can be made between the new measures in question and the standard test that the investigator wishes to elucidate. Componential approaches wherein attempts are made to understand a complex task by combining latencies or other measures from sets of simpler tasks, particularly when the method of precuing is used, fare worst in this regard. Difficulties arise when trying to build a case that sets of simple tasks provide an equivalent task environment to the task demands of the complex task the investigators are seeking to understand. In contrast, componential approaches making use of intact tasks fare quite well because the task under investigation is not altered.

Ease of Implementation and Scoring

The computerized testing technologies fare well along this dimension. Once the considerable time and expense for set-up is overcome, it is possible to collect and score data on large numbers of subjects with relatively little human intervention, compared to examiner-based testing. Psychophysiological and electrophysiological measurements requiring expensive equipment and/or specially trained examiners make implementation slightly more complicated, although scoring typically is computerized. The uniqueness and richness of the data collected by these techniques seems to entirely justify added time and expense. Measures yielding writing samples, such as the MCAT project, do not pose any difficulty in terms of administration, although achieving reliable scoring and establishing valid competency standards provide a worthwhile challenge.

Replicability/Standardizability

Computerized and equipment-based testing approaches have the obvious advantage of providing standardizable testing sessions. Dynamic testing techiniques, where examiners' questions or responses may depend on subjects' responses, may pose a challenge in this regard. Problems can be avoided by creating templates of questions and responses for various types of subject responses prior to testing, based on pilot data. Again, the richness of the information available through dynamic testing clearly justifies this additional attention.

Criterion-Related Validity

Certainly, investigators are concerned with the extent to which particular aptitude testing approaches account for significant amounts of variance in measures of important types of intelligent behavior. Where predictive validity data are provided in the work reported in this volume, estimates are impressive. Information processing measures have the potential to tap fundamental information processes common to predictor and criterion tasks (Dillon, 1986) and, thus, the measures offer great promise for improving efforts to predict academic and occupational success. By the same token, improving the way investigators measure familiar aptitudes and extending those aptitude constructs to include more realistic measures of intelligent performance move researchers toward the goal of improved prediction of intelligent performance. Finally, conceptualizing new aptitudes, such as tacitly acquired knowledge and other measures of practical intelligence, as well as aptitudes such as cognitive flexibility, are promising in that they appear to measure heretofore untapped variance in criterion task performance.

SUMMARY

In this chapter, I have described and comparatively evaluated a set of new methods and measures for assessing intellectual skills. New methods of measuring these predictors of intelligent performance include dynamic testing procedures and varieties of computerized test administration. New measures taken from aptitude tests include information processing componential and metacomponential skills, tacitly acquired knowledge, cognitive flexibility, and affective variables. Approaches were evaluated with respect to their theoretical grounding, psychodiagnostic utility, the ease of administration and scoring, replicability/standardizability, and criterion-related validity.

New work in criterion development also was considered. The use of comprehensive clinical competency exams in areas such as medical education reflect a trend toward standardization of criterion data as well as a move toward using criterion measures that more realistically reflect competencies being trained. Other new progress includes the use of tutor labs to provide criterion data that are more easily obtainable than other criterion assessments.

Exciting progress has been made in the past ten years with regard to development of theoretically motivated testing programs. This work reflects new developments in knowledge regarding the nature and organization of intellectual abilities, and the work provides fertile ground for continuing theoretical and applied advances.

REFERENCES

Anderson, S. B., and Dillon, R. F. (1987). *Predicting academic performance among low performing college freshmen*. Unpublished manuscript.

Appel, V., and Watson, T. (1987). *Officer selection*. Unpublished manuscript.

Association of American Medical Colleges. (1984). *The MCAT Student Manual*. Washington, D.C.: Author.

Carlson, J. C., and Dillon, R. F. (1979a). Effects of testing conditions on Piaget matrices and order of appearance problems: A study of competence versus performance. *Journal of Educational Measurement, 16*(1), 19–26.

_____ (1979b). Assessment of intellectual capabilities in hearing impaired children: Effects of testing-the-limits procedures. *Volta Review, 8*(4), 216–224.

Copper, L. A., and Mumaw, R. J. (1985). Spatial aptitude. In R. F. Dillon (Ed.), *Individual differences in cognition* (Vol. 2). New York: Academic Press.

Dillon, R. F. (1979). Testing for competence: Refinement of a paradigm and its application to the hearing impaired. *Educational and Psychological Measurement, 39*, 363–371.

_____ (1981a). Analogical reasoning under different methods of test administration. *Applied Psychological Measurement, 5*(3), 341-347.

_____ (1981b). *Individual differences in eye fixations within and between stages of inductive reasoning* (Report No. 81–2). Carbondale, IL: Southern Illinois University, Eye Movement Research Laboratory.

_____ (1985a). Predicting academic achievement with models based on eye movement data. *Journal of Psychoeducational Assessment, 3*, 157–165.

_____ (1985b). Eye movement analysis of information processing under different testing conditions. *Contemporary Educational Psychology, 10*, 387–395.

_____ (19886). Information processing and testing. *Educational Psychologist, 21*(3), 163–174.

_____ (1987a). *An information-processing framework for understanding intelligent performance*. Manuscript submitted for publication.

_____ (1987b). *Metacomponents of intelligence*. Manuscript submitted for publication.

_____ (1987c). Information processing and intelligence. In R. J. Sternberg (Ed.), *Advances in the psychology of human intelligence* (Vol. 5). Hillsdale, NJ: Erlbaum.

_____ (1987d). *Flexibility as a measure of intelligent performance*. Manuscript in preparation.

Dillon, R. F., and Carlson, J. (1978). Testing for competence in three ethnic groups. *Educational and Psychological Measurement, 38*(2), 437–443.

Dillon, R. F., and Donow, C. (1982). The psychometric credibility of the Zelinder and Jeffrey modification of the Matching Familiar Figures Test. *Education and Psychological Measurement, 42*, 529–536.

Dillon, R. F., Loschen, E., and Travis, T. (1987). *Flexibility as a measure of intelligence among psychiatry clerks.* Unpublished manuscript.

Dillon, R.F., and Reznick, R. K. (1986). Development of a new system of measurement. In *Proceedings of the 1986 Meeting of the Military Testing Association.* Mystic, CN: Military Testing Association.

Dillon, R. F., Wagner, R., Loschen, E., and Travis, T. (1987). *Tacit occupational knowledge in psychiatry.* Manuscript in preparation.

Embretson, S.E. (1983). Construct validity: Construct representation and nomothetic span. *Psychological Bulletin, 93*, 179-197.

_____ (1984). A general latent trait model for response processes. *Psychometrika, 49*, 175–186.

_____ (1985). Multicomponent latent trait models for test design. In S. E. Embretson (Ed.), *Test design: Developments in psychology and psychometrics.* New York: Academic Press.

_____ (1986). *The effect of dynamic testing on validity.* Paper presented at the Annual Convention of the American Educational Research Association, San Francisco. CA.

_____ (in press). The psychometrics of dynamic testing. In C. Lidz, *Dynamic testing.* Beverly Hills, CA: Guilford Press.

Feuerstein, R. (1979). *The dynamic assessment of retarded performers*: *The learning potential assessment device, theory, instruments, and techniques.* Baltimore: University Park Press.

Hunt, E. B. (1983). On the nature of intelligence. *Science, 218*(4581), 141–146.

Jencks, C. (1977). *Who gets ahead?* New York: Basic Books.

Kyllonen, P. C. (1986). Theory-based cognitive assessment. In J. Zeidner (Ed.), *Human productivity enhancements*: *Organizations, personnel, and decision making* (Vol. 1, pp. 338–381). New York: Praeger.

Lewis, G. W. (1983). Event related brain electrical and magnetic activity: Toward predicting on-job performance. *International Journal of Neuroscience, 18*, 159–182.

Lewis, G. W., Trejo, L. J., Blackburn, M. R., and Blankenship, M. H. (1986). Neuroelectric and neuromagnetic recordings: Possible new predictors of on-job performance. In G. E. Lee (Ed.), *Psychology in the Department of Defense* (Tenth Symposium). Colorado Springs: U.S. Air Force Academy (USAFA TR 86-1).

Matarazzo, J. D. (1972). *Wechsler's measurement and appraisal of adult intelligence* (5th ed.). Baltimore: Williams and Wilkins.

McCall, R. B. (1977, July 29). Childhood IQ's as predictors of adult educational and occupational status. *Science, 197*, 482-483.

Molfese, D. L., and Molfese, V. J. (1985). Predicting a child's preschool language performance from perinatal variables. In R. F. Dillon (Ed.), *Individual differences in cognition* (Vol. 2). New York: Academic Press.

Mullis, I. V. S. (1985). *NAEP perspectives on literacy: A preview of 1983-1984 writing assessment results, the young adult literacy assessment and plans for*

1986. Paper presented at the Annual Convention of the American Educational Research Association, Chicago, IL.

Pellegrino, J. W., and Goldman, S. R. (1983). Developmental and individual differences in verbal and spatial reasoning. In R. F. Dillon and R. R. Schmeck (Eds.), *Individual differences in cognition* (Vol. 1). New York: Academic Press.

Snow, R. E., and Lohman, D. F. (1984). Toward a theory of cognitive aptitude for learning from instruction. *Journal of Educational Psychology, 76*(3), 347-376.

Sternberg, R. J. (1981). What should intelligence tests test? Implications from a triarchic theory of intelligence for intelligence testing. *Educational Researcher, 13*(1), 5-15.

―――― (1985). *Beyond IQ: A triarchic theory of human intelligence.* Cambridge: Cambridge University Press.

Sternberg, R. J., and Wagner, R. K. (1986). *Practical intelligence: Origins of competence in the everyday world.* Cambridge: Cambridge University Press.

Vygotsky, L. S. (1978). *Mind in society: the development of higher psychological processes.* Cambridge: Harvard University Press.

Wechsler, D. (1975, January). Intelligence defined and undefined: A relativistic appraisal. *American Psychologist, 30*(2), 135-139.

Williams, R. (1987). *Criterion development in medical education.* Unpublished manuscript.

2   An Overview of the Armed Services Vocational Aptitude Battery (ASVAB)

Paul Foley
Linda S. Rucker

INTRODUCTION

The Armed Services Vocational Aptitude Battery (ASVAB) is a joint service battery that, in conjunction with other criteria, is used for the selection of applicants for enlistment into the Armed Forces and the classification of those accepted as recruits. The ASVAB is a paper and pencil test that requires approximately 3 hours to administer. Chapter 4 entitled "Computerized Adaptive Testing of a Vocational Aptitude Battery," describes efforts directed toward the computerized administration of the ASVAB.

The Armed Forces Qualification Test (AFQT) formed of four of the ten ASVAB subtests, is the selection instrument used to determine an applicant's eligibility for enlistment. Applicants qualified for enlistment are classified into a military occupation by using various composites of ASVAB subtests that predict success in initial occupational training schools.

In fiscal year 1985, the ASVAB was administered to approximately 1.5 million high school students and recruit applicants, which made it the most widely used multiple aptitude battery in the United States. The following review is intended to provide a background for researchers unfamiliar with the ASVAB.

BACKGROUND

U.S. armed service aptitude testing originated during World War I when the Army Alpha and Beta tests were developed for evaluating large groups of recruits. Stuit (1947) reported that the first recorded organized testing program of naval personnel was in 1924 when the Training Division was established. The General Classification Test was used at that time for the selection of recruits for Navy schools. During World War II, the Army General Classification Test (AGCT) and the Navy General Classification Test (NGCT) were used for selection. The U.S. Navy used the Basic Test Battery (BTB) for the classification of recruits into specific occupations (ratings).

Subsequently, use of a common selection instrument (based upon a World War II reference population) throughout the military was mandated by passage of the Selective Service Act of 1948. In 1950, the AFQT was implemented as the selection instrument to determine which Selective Service registrants and volunteer applicants were qualified to enter military service. Although the services used a common AFQT for selecting recruits, they used their own battery for occupational classification. Between 1973 and 1975, the services stopped using a common AFQT and developed their own conversion tables for estimating individual AFQT scores from their own occupational classification batteries.

In 1968, the Department of Defense (DoD) introduced the ASVAB, the first joint service test battery, as part of the Student Testing Program where it was offered to student in grades 11 and 12 and postsecondary schools. Each service adopted the ASVAB in 1976 for its Production Testing Program and for selecting enlistees and classifying them into military occupations.

The reasons for adopting a joint service battery for the Production Testing Program were to: (1) avoid subjecting persons applying to more than one service to multiple testing; (2) facilitate inter-service referrals of applicants; and (3) enable service psychologists to focus their efforts on a single enlisted selection and classification battery. The role of the Air Force, which had been the executive agent for the DoD Student Testing Program was expanded to include the Production Testing Program. The Air Force became reponsible for the research and development of both programs (ASVAB Working Group, 1980).

TEST FORMS AND SUBTEST DESCRIPTIONS

New forms of the ASVAB used for production testing have been introduced approximately every four years. This routine updating is intended to improve the security of tests, to replace obsolete test items, and to incorporate improvements in the field of psychological measurement into the new tests. This section presents: (1) the history of the ASVAB in the DoD Student Testing Program, (2) the history of the ASVAB in the Production Testing Program, and (3) a description of the subtests in the current ASVAB.

DoD Student Testing Program

ASVAB Form 1 was introduced to the DoD Student Testing Program in 1968, and replaced by ASVAB Form 2 in 1973.[1] ASVAB Form 5 was introduced in 1976 and replaced by ASVAB 14 in July 1984.[2]

Production Testing Program

ASVAB Forms 6 and 7 were introduced in 1976, the first year the Joint Service Production Testing Program used the ASVAB. Each ASVAB form has a version A and B. The items within the AFQT subtests for version A are different from version B. The items within the non-AFQT subtests of versions A and B are identical but their order within each subtest is scrambled. ASVAB Forms 6 and 7 were similar to Form 5 but contained the Army Classification Inventory, which was appropriate to the Production Testing Program but unnecessary for the DoD Student Testing Program.

ASVAB Forms 8, 9 and 10 replaced Forms 6 and 7 in October 1980. They included Paragraph Comprehension (PC) and Coding Speed (CS) subtests as well as a combination of Automotive Information and Shop Information subtests of Forms 6 and 7 called the Auto and Shop Information (AS) subtest, but did not include the Space Perception (SP), Attention to Detail (AD), General Information (GI) subtests or the Army Classification Inventory.

This reduced the total number of ASVAB subtests from 16 to 10 and the number of items by 10 percent. ASVAB Forms 8, 9, and 10 were replaced by parallel ASVAB Forms 11, 12, and 13 in October 1984.

ASVAB Subtests

ASVAB Form 14 is now being used with the DoD Student Testing Program, while ASVAB Forms 11, 12, and 13 are being used with the Production Testing Program. Forms 11–14 (as well as 8–10) are comprised of ten separately timed subtests; eight are power tests and two are speeded tests. The major differences between the power and speeded tests concern difficulty level of test items and time limits for completion. Items on the power tests span a broad range of difficulty—some can be answered accurately by nearly all examinees and others by very few. However, most examinees are able to finish these tests in the time allotted. Conversely, speeded items are easy for all examinees to answer correctly if given sufficient time.

The following subtests are classified as power test: General Science (GS), Arithmetic Reasoning (AR), Word Knowledge (WK), Paragraph Comprehension (PC), Auto and Shop Knowledge (AS), Mathematics Knowledge (MK), Mechanical Comprehension (MC), and Electronics Information (EI). The speeded tests are Numerical Operations (NO) and Coding Speed (CS). Brief descriptions of the subtests are provided in Table 2.1.

DEVELOPMENT OF ASVAB FORMS 11, 12, AND 13

This section discusses the development of the ASVAB Forms 11, 12, and 13 which are currently in use.

Pretesting of Test Items

Development of ASVAB Forms 11, 12, and 13 began with development of item pools for each subtest. Table 2.2 shows the item taxonomy used for development of parallel forms of the ASVAB. This item taxonomy, revised in 1986, provided explicit rules for the development of forms similar to ASVAB Form 8 (version A). Form 8A is referred to as the anchor because subsequent test forms must be parallel to it to be considered equivalent.

The items were assembled into test booklets that contain more questions per test than necessary. After the preliminary tests were administered to Air Force recruits, the traditional discrimination indexes (biserial and point-biserial correlations) of each item as well as the difficulty index (proportion of examinees responding correctly) were determined. The item response theory (IRT) parameters of (1) discrimination, (2) difficulty, and (3) guessing were also determined. Item analysis information was used to assemble six test batteries equivalent to one another and to previous ASVAB forms.

Equating

To be meaningful, an examinee's raw test score must be interpreted with reference to a normative group composed of a large number of subjects

Table 2.1
A Description of ASVAB Tests

Test	Abbreviation	Description
General Science	GS	25 items measuring general knowledge of the biological and physical sciences, administered with an 11-minute time limit.
Arithmetic Reasoning	AR	20 items that require examinees to solve word problems that typically involve simple calculation. This test is administered with a 36-minute time limit.
Word Knowledge	WK	35 items that require examinees to select the correct meaning of the word, or to identify a synonym. Time limit is 11 minutes.
Paragraph Comprehension	PG	Examinees are required to read several short paragraphs and to answer 15 questions that assess their understanding of what they have read. Time limit is 13 minutes.
Numerical Operations	NO	Examinees are given 3 minutes to solve 50 items involving simple calculations. This test is designed to measure speed of calculation.
Coding Speed	CS	A highly speeded coding task in which examinees are given 7 minutes to complete 84 items. This test requires examinees to substitute numeric codes for verbal material.
Auto and Shop Knowledge	AS	25 items measuring knowledge of automobiles, tools, shop terminology, and shop practices, administered with 11-minute time limit.
Mathematics Knowledge	MK	25 items measuring knowledge of high school level mathematics (algebra, geometry, elementary trigonometry), administered with a 24-minute time limit.
Mechanical Comprehension	MC	25 items measuring knowledge of basic mechanical and physical principles, administered with a 19-minute time limit. Several pictorial items, similar to those that make up the Bennett Test of Mechanical Comprehension are included.
Electronics Information	EI	20 items measuring knowledge of electrical principles and electronic terminology, administered with a 19-minute time limit.

Table 2.2
ASVAB Item Taxonomy

Test	Content or Skill
General Science (GS)	**Content** • Life Science • Physical Science • Earth Science
Arithmetic Reasoning (AR)	**Skill** • Recognition and application of the four basic arithmetic operations • Rearrangement of these operations • Computations of percentages • Solution of "rate" problems • Conversion of simple units of time and distance • Computation of perimeters, areas, and volumes of geometric forms (Note. Items include one or more of these operations.)
Word Knowledge (WK)	**Skill** • Identification of synonyms within the context of sentences
Paragraph Comprehension (PC)	**Skill** • Recall detail • Paraphrase or summarize text • Derive inferences from text • Generalize knowledge to novel situations • Recognize and comprehend sequential, cause/effect, and comparative relationships
Numerical Operations (NO)	**Skills** • Perform with speed problems involving the four arithmetic operations
Coding Speed (CS)	**Skill** • Perform with speed matching of 4-digit numbers with single words where placement of words is determined randomly
Automotive and Shop Information (AS)	**Content** • Engines, body/drive train, and electronics (automotive) • Tools and materials (shop)
Mathematical Knowledge (MK)	**Skills** • Conversion of common fractions, decimals, and percents • Simplification of fractions, improper fractions, and reciprocals • Determining least common denominators, greatest common factors, and smallest common multiples • Analytic and plane/solid geometry • Operations of exponents, roots, powers, and polynomials • Equation solving • Transformation of verbal problems into algebraic symbols
Mechanical Comprehension (MC)	**Content** • Simple machines • Basic compound machines • Complex compound machines • Structural components • Mechanical concepts
Electronics Information (EI)	**Content** • Theory and principles • Circuit diagnosis • Power and electricity • Tools and regul

from the target population who had been administered the same test. The norm group for ASVAB Forms 11, 12, and 13 is the 1980 American Youth Population that had been administered the anchor ASVAB Form 8A. To use the 1980 American Youth Population for interpretation of the six new ASVABs (11A and B, 12A and B, and 13A and B), it was necessary to equate these new forms to the anchor Form 8A. This topic is discussed more thoroughly later in the chapter on reference populations and norming the ASVAB.

Versions A and B of ASVAB 11, 12, and 13 were equated to Form 8A by using an equivalent groups design. During the period January to March 1983, 15,000 service recruits were examined on either ASVAB 8A or one of the new ASVABs 11, 12, or 13 (versions A and B) randomly assigned. During the same period, nine partial batteries constructed from ASVAB Form 11A and nine partial batteries constructed from ASVAB Form 8A were administered to 78,000 service applicants, again using an equivalent groups design. The new forms were equated to the anchor ASVAB 8A, and a linear equating table for five of the six ASVAB forms was developed. Form 12A was considered to be less parallel to the other forms and a separate equating table was developed. For more information on development and equating of ASVAB Forms 11, 12, and 13, see Prestwood, Vale, Massey, and Welsh (1985) and Ree, Welsh, Wegner, and Earles (1985).

Item Analysis

All of the subtests in the six new ASVABs are parallel to one another. Table 2.3 shows the parallelism between three of the six new ASVABs (Forms 11A, 12B, and 13A) and the anchor ASVAB 8A. Table 2.3 contains the mean of the traditional item discrimination index and the item difficulty index for each of the ten subtests. Similar information is provided for the three item response theory (IRT) parameters for all the subtests except the two speeded subtests (NO and CS).

Reliability

The Kuder-Richardson Formula 20 (KR-20) reliabilities are provided in Table 2.4 for the power subtests of ASVAB Forms 11A, 12B, 13A, and 8A (the same forms as on Table 2.3). Approximately 2,000 examinees' test results were used in determining the reliability of each test. The reliability of the two speeded subtests cannot be determined directly using the KR-20 formula; however, the reliability of these two speeded tests has been conservatively estimated to be .70 (Department of Defense, 1984a).

Table 2.3
Mean Item Statistics for ASVAB Forms 11A, 12B, 13A, and 8A

| Test | Form | Classical | | Item Response Theory | | |
		Discrimination	Difficulty	Discrimination	Difficulty	Guessing
GS	11a	.618	.686	1.147	-0.496	.200
	12b	.607	.685	1.161	-0.571	.209
	13a	.610	.688	1.204	-0.470	.218
	8a	.549	.679	1.043	-0.496	.217
AR	11a	.629	.644	1.125	-0.321	.189
	12b	.626	.642	1.179	-0.265	.196
	13a	.593	.646	1.125	-0.320	.203
	8a	.611	.607	1.208	-0.223	.186
WK	11a	.705	.766	1.244	-0.817	.237
	12b	.717	.765	1.341	-0.757	.243
	13a	.687	.760	1.302	-0.757	.249
	8a	.667	.785	1.240	-1.090	.245
PC	11a	.695	.741	1.271	-0.743	.204
	12b	.725	.776	1.252	-0.837	.229
	13a	.695	.756	1.379	-0.700	.249
	8a	.648	.745	1.150	-0.627	.249
NO	11a	.711	.718	--a	--a	--a
	12b	.709	.691			
	13a	.704	.712			
	8a	.687	.727			
CS	11a	.769	.560	--a	--a	--a
	12b	.770	.563			
	13a	.772	.566			
	8a	.778	.563			
AS	11a	.620	.662	1.162	-0.342	.192
	12b	.621	.636	1.219	-0.173	.195
	13a	.622	.668	1.111	-0.385	.195
	8a	.577	.653	1.044	-0.311	.216
MK	11a	.607	.529	1.256	0.211	.154
	12b	.661	.513	1.332	0.263	.146
	13a	.597	.530	1.161	0.179	.157
	8a	.590	.531	1.221	0.137	.162
MC	11a	.564	.623	0.902	-0.219	.197
	12b	.571	.606	0.946	-0.142	.189
	13a	.567	.626	0.968	-0.182	.207
	8a	.573	.593	0.976	-0.061	.186
EI	11a	.584	.605	1.184	-0.070	.189
	12b	.571	.640	.988	-0.268	.207
	13a	.574	.630	1.150	-0.105	.214
	8a	.567	.625	1.067	-0.134	.195

Note. The classical discrimination index is the biserial correlation.

[a]Parameters were not developed for the two speeded tests, NO and CS, because no adequate procedure exists for these parameters. For further explanation, see Birnbaum (1968).

Table 2.4
Reliability (KR-20)

Test	Forms			
	11a	12b	13a	8a
GS	.824	.812	.820	.769
AR	.881	.877	.859	.877
WK	.892	.893	.885	.864
PC	.780	.765	.754	.722
AS	.850	.854	.851	.824
MK	.854	.884	.847	.842
MC	.814	.821	.813	.826
EI	.783	.773	.767	.760

SELECTION OF APPLICANTS

The role of the AFQT is to provide a measure of general trainability of applicants accepted for enlistment into the services. The AFQT for ASVAB Forms 5, 6, and 7 consisted of the WK, AR, and SP subtests. The AFQT for Forms 8 through 14 consisted of the Verbal (VE) (a combination of the WK and PC subtests), the AR subtest, and one half of the NO subtest. The sum of raw scores on these subtests is converted to a percentile score.

With introduction of ASVAB Forms 15, 16, and 17, the composition of the AFQT as well as the method of combining the scores on the individual subtests will change. The MK subtest will replace the one half of the NO subtest. The raw scores on the VE, AR, and MK will be converted to standard scores. VE will be double weighted and added to the sum of the AR and MK converted scores. Finally, the sum of the standard scores will be converted to a percentile score.

The services now use the minimum qualifying scores:

1. *Army.* An AFQT score of 16 with a high school diploma, or 31 with a General Education Development (GED) high school equivalency or with no diploma.
2. *Navy.* An AFQT score of 17 with a high school diploma, 31 with a GED, or 38 without a high school diploma.
3. *Marine Corps.* An AFQT score of 21 with a high school diploma, 31 for other credential holders, and 50 without a high school diploma.
4. *Air Force.* An AFQT score of 21 with a high school diploma, 50 with a GED, or 65 without a high school diploma.

Table 2.5
AFQT Categories of Percentile Scores

Category	Percentile Score Range
I	93 and above
II	65-92
IIIa	50-64
IIIb	31-49
IV	10-30
V	1-9

Minimum qualifying AFQT percentile scores are determined by considering the manpower needs in terms of quality and quantity and the status of the recruiting market (i.e., the availability of the armed services to meet its manpower needs from the available supply of qualified youth). As these factors change periodically so will the minimum qualifying scores. For reporting purposes, the percentile AFQT scores are grouped into 6 broad catagories, as shown in Table 2.5.

In addition to determining recruit quality through AFQT categorization, educational level is included in the assessment process. The armed services place a high premium on applicants who have a high school diploma because this tends to be the best single predictor of adaptability to military life (Eitelburg, Laurence, Waters, & Perelman, 1984).

OCCUPATIONAL CLASSIFICATION

ASVAB Composites

Classification of applicants into military occupations is accomplished by using various composies of ASVAB subtests that predict success in the associated initial occupational training school. To qualify for entry into a specific military occupation, an applicant must receive the minimum qualifying ASVAB composite score for the training school. In addition to considering the applicant's qualifications, the classification process must balance the applicant's personal preferences and the needs of the service before making a final assignment. The composites are routinely revalidated by each of the services and changes to the minimum qualifying scores made as required.[3]

For the Production Testing Program, the Army structured the subtests into nine composites; the Navy, 11 composites; and the Air Force and the Marine Corps, four composites. The DoD Student Testing Program has the following seven composites: Academic Ability; Verbal; Mathematics; Mechanical and Crafts; Business and Clerical; Electronics and Electrical; and Health, Social, and Technology. Table 2.6 lists the subtests in each composite.

Table 2.6
Operational Composites Currently Used by the Various Services

Test	Composites
Army	
Clerical/Administrative	VE + AR + MK
Combat	AR + CS + AS + MC
Electronics Repair	GS + AR + MK + EI
Field Artillery	AR + CS + MK + MC
General Maintenance	GS + AS + MK + EI
Motor Maintenance	NO + AS + MC + EI
Operators/Food⁻	VE + AR + AS + MC
Surveillance/Communications	VE + NO + CS + AS
Skilled Technical	GS + VE + MK + MC
Navy	
General Technical	VE + AR
Mechanical	VE + MC + AS
Electronics	AR + MK + EI + GS
Clerical	VE + NO + CS
Basic Electricity and Electronics	AR + 2MK + GS
Engineering	MK + AS
Communications Technician	VE + AR + NO + CS
Hospitalman	VE + MK + GS
Machinery Repairman	AR + MC + AS
Submarine	VE + AR + MC
Business and Clerical	VE + MK + CS
Air Force	
Mechanical	GS + 2AS + MC
Administrative	VE = NO + CS
General	AR + VE
Electronic	AR + MK + GS + EI
Marine Corps	
Mechanical Maintenance	AR + AS + MC + EI
Clerical	VE + MK + CS
Electronics Repair	GS + AR + MK + EI
General Technical	VE + AR + MC
Student Testing Program	
Academic Ability	AR + VE
Verbal	GS + VE
Mathematics	AR + MK
Mechanical and Crafts	AR + AS + MC + EI
Business and Clerical	VE + CS + MK
Electronics and Electrical	GS + AR + MK + EI
Health, Social, and Technology	AR + VE + MC

Note. VE is a combination of the WK and PC tests.

Composites are computed by converting raw test scores to standard scores with a mean of 50 and standard deviation of 10. The tests' means and standard deviations were obtained from the 1980 American Youth Population. The Navy's composites are produced by summing the standard scores across tests. The Army and Marine Corps composites are produced by summing the standard scores across tests and transforming the sum to a mean of 100 and a standard deviation of 20. The Air Force composites are produced by summing the standard scores across tests and converting these sums to percentiles. The composites used in the high school testing program are produced by summing the standard scores across tests and transforming them to composite standard scores with a mean of 50 and a standard deviation of 10. The composite standard scores are converted into percentile scores based upon normative grade and gender subgroups (Maier & Truss, 1985).

Validity

The ASVAB composites used in the Production Testing Program are validated against final grades received at completion of entry-level occupational training. Table 2.7 contains sample validity information from the *Test Manual for the ASVAB* (Department of Defense, 1984a). The missing validity coefficients were not reported in the original validity study.

For additional predictive validity on ASVAB composites used by the Army, see McLaughlin, Rossmeissl, Wise, Brandt, and Wang (1984); the Navy, Booth-Kewley, Foley, and Swanson (1984); and the Air Force, Wilbourn, Valentine, and Ree (1984) and Maier and Truss (1985).

Each of the services has traditionally validated ASVAB composites against end-of-training grades. In 1980, however, a number of projects were initiated to develop measures of job performance that could be used for

Table 2.7
Validities of Service Composites

Service	Specialty Title	Sample Size	Validity	Composite
Army	Signal/Security Specialists	91	48/79	Surveillance/ Communications
	Legal Clerk	96	27/64	Clerical
Navy	Cryptologic Technician	140	50/59	General Technical
	Equipment Operator	181	22/36	Mechanical
Air Force	Avionics Sensor System Specialist	244	49/—	Electronics
	Helicopter Mechanic	155	46/—	Mechanical
Marine Corps	Hawk Missile System Operator	107	—/31	General Technical
	Field Radio Operator	903	—/47	Electronics

Note. Validities to the left of the slash are uncorrected for restriction in range, those to the right have been corrected.

ASVAB validation. This work was prompted by a congressional mandate to link military selection standards on-the-job performance rather than to the end-of-training grades. The various efforts are described and assessed in Wigdor and Green (1986).

ASVAB and Automated Classification

In 1981, the Navy began using an automated classification system called CLASP (Classification and Assignment within PRIDE-Personalized Recruiting for Immediate and Delayed Enlistment). CLASP is modeled after the Air Force PROMIS (Procurement Management Information System) that is an integral part of the classification interview. It matches an applicant with optimal Navy jobs by simultaneously considering the needs of the Navy and the needs of the applicant. CLASP has five components that are designed to: (1) maximize the probability of an individual completing Class "A" school as based upon ASVAB test scores, (2) provide the best match between an individual's ability as measured by ASVAB scores and the requirements of a job, (3) accommodate both an applicant's occupational preference and the Navy's manpower goals, (4) fill all Class "A" school seats at a uniform rate, and (5) consider ethnic minorities for assignment to all ratings to ensure representative composition. CLASP provides a list of Navy ratings for which the applicant is qualified. With the assistance of the classifier, the applicant selects the Class "A" school he/she will attend upon completion of boot camp. A description of the CLASP-PRIDE system can be found in Kroeker and Rafacz (1983).

Each service uses its own unique computer model, that reflects its current standards, policies, and relative priorities, for the assignment of recruits to specific training schools and jobs. The Army's Recruiting Quota System (REQUEST) and the Air Force's PROMIS are similar to the Navy CLASP system, while the Marine Corps uses a recruit distribution system that assigns recruits to the most demanding job opening that exists at a given time for which they are qualified. However, minority quotas and scheduling of training classes also help to determine these assignments.

REFERENCE POPULATIONS AND NORMING THE ASVAB

World War II Reference Population

Since the ASVAB was introduced as a joint service instrument, two reference populations have been used. The World War II reference population was used from 1950 to October 1984 when it was replaced by the 1980 American Youth Population. The World War II reference population referred to the extensive testing of adult males during World War II on the Army General Classification Test (AGCT) and the Navy General Classification Test (NGCT), that were later replaced by the AFQT.[4] The AFQT was introduced operationally on January 1, 1950, with Forms 1 and 2, followed by Forms 3 and 4 in January 1953, Forms 5 and 6 in August 1956, and Forms 7 and 8 in July 1960. At that time, the AFQT was a separate test and not a composite of subtests from a larger battery as it is today with the ASVAB. Each new form of the AFQT was administered in addition to either AGCT Form 1C or a previous AFQT and the two test score distributions were equated so that the scores on the new AFQT would have the same meaning in terms of trainability as scores on the original AGCT.

Some of the advantages of using the World War II reference population were:

1. The performance level associated with test scores was corroborated by the abundant and well-documented research collected during World War II; hence, enlistment standards and prerequisite scores for assignment to training schools could be confidently established.
2. The scale score accurately described the aptitudes of the mobilization population during the early 1950s, that assisted manpower planning for the Korean War (Department of Defense , 1986).[5]

The services were not required to use the AFQT from 1973 through 1975. Instead, they were permitted to develop conversion tables from their own service battery as an estimate of an individual's AFQT score. With introduction of ASVAB in 1976, each of the services again used a common AFQT formed of three ASVAB subtests. The ASVAB AFQT, however, was not normed back to the AGCT as the preceding AFQT Forms 1 through 8 had been but was normed against the AFQT scores derived by the services from different batteries (Army, AFQT from Army Classification Battery (ACB–73) and Navy and Air Force, ASVAB Form 2). After implementation, an excessive number of persons were found to be scoring in the upper two mental categories of the AFQT and an adjustment was made that reduced the numbers in these two categories. Later, it was found that many individuals scoring in Category III should have been placed in Category IV. In October 1980, the problem of miscalibration associated with ASVAB Forms 6 and 7 was corrected with introduction of ASVAB Forms 8, 9, and 10 whose scores had been properly equated to the World War II reference population. For more about the ASVAB misnorming, see ASVAB Working Group (1980) and Maier and Truss (1983).

1980 American Youth Population

The change from the World War II reference population to the 1980 American Youth Population was in response to: (1) major educational and cultural changes that occurred in the general population during the 1960s and 1970s; (2) the fact that World War II reference population consisted only of males; and (3) the lack of insight that comparing current enlistees with a World War II reference population provides into enlistees' aptitude ranking in relation to their contemporary, civilian counterparts.

Therefore, the services decided that a new reference population was essential. The Department of Defense (DoD) in conjunction with the

Department of Labor, sponsored the Profile of American Youth Study. Its goals were to: (1) assess the vocational aptitudes of persons aged 16 to 23; (2) establish a reference population against which scores on DoD enlistment tests could be interpreted; and (3) compare recruit aptitudes with those of the youth population in general. In 1980, under contract from DoD, the National Opinion Research Center (NORC) of the University of Chicago tested approximately 12,000 youths using ASVAB Form 8A. This sample had approximately equal proportions of male and female respondents from all major U.S. census regions. Certain subgroups of interest (Hispanics, blacks, economically disadvantaged whites, and women in the military) were oversampled to assure representative groups of sufficient size for valid subgroup comparisons. These samples were then statistically weighted to represent the population demographics of American youth. The resulting sample is known as the American Youth Population. A subset of the 18–23-year-old males and females in this sample is the new reference population for norming ASVAB scores from Military Entrance Processing Stations (MEPs) since the services typically only recruit young people 18 years of age and older. High school norms are based on eleventh and twelfth grade reference populations. The eleventh grade reference population of 1,304 persons and the twelfth grade reference population of 1,253 persons were both obtained from the 1980 American Youth Population (Bock & Moore, 1984). Construction of the ASVAB scale scores is reported by Maier and Sims (1986). Recently, Divgi and Horne (1986) investigated the appropriateness of using the ASVAB with the ninth and tenth grades and developed norms for these grades. Only the tenth grade norms, however, were implemented in July 1986.

Referencing of ASVAB to the 1980 American Youth Population commenced with the introduction of Forms 11, 12, and 13 in October 1984. The procedure for referencing Forms 11, 12, and 13 to ASVAB Form 8A in the 1980 American Youth Population was covered in the section on Equating.

Implementation of ASVAB Forms 11, 12, and 13 was delayed as a result of the findings of Sims and Maier (1983). They discovered notable differences between males of the 1980 American Youth Population and samples of male military applicants and recruits on the Numerical Operations (NO) and Coding Speed (CS) subtests of ASVAB Form 8A. Scores for the military sample were estimated as three raw score points higher on NO and one raw score point higher on CS. Earles, Giuliano, Ree, and Valentine (1983) attributed this difference to the different test answer sheets used by the respective groups. Wegner and Ree (1985) developed corrected conversion tables for the NO and CS tests and Forms 11, 12, and 13 were finally introduced in October 1984.

An Initial Operational Test and Evaluation (IOT&E) was conducted with introduction of ASVAB 11, 12, and 13 to assure the correctness of the conversion tables. IOT&E used an equivalent groups design in which either the anchor test (Form 8A) or one of the six new forms was randomly assigned to service applicants. The IOT&E period was originally scheduled for two months but was extended to four months. During IOT&E, the mean AFQT score on Form 8A (relabeled 13C) was found to be higher than the mean AFQT score obtained from the six new test forms. The problem was traced to the NO subtest, which is one of four subtests used in computing AFQT. The source of the problem was a difference between the print format of Form 13C and that of the six new forms. This was confirmed by administering the NO subtest in the Form 13C print format to a group of 120 Air Force recruits and the NO subtest in the Forms 11, 12, and 13 print format to another group of 120. The group administered NO in the Form 13C received a mean score approximately one point higher than did the other groups (Department of Defense, 1986).

The print format for Form 13C was more open that that of Forms 11, 12, and 13. This facilitated rapid responses which is important because NO is a speeded test. The print format for Form 13C in the IOT&E was the same as Form 8A in the 1980 American Youth Population. It was therefore concluded appropriate to use the IOT&E data to equate the new test forms to the anchor test Form 8A used in the American Youth Population (Department of Defense, 1986).

INDEPENDENT REVIEWS OF THE ASVAB

A number of researchers have assessed the ASVAB. Cronbach's (1979) review of ASVAB Form 5 was critical of its use as a tool for vocational counseling of high school students. His main criticism was that the six composites making up the profile supplied to students in 1976–77 school year were highly correlated and provided virtually no differential prediction for occupational specialties. Cronbach was also concerned with: (1) the difficulty of some tests (for females, MC and AS); (2) the inadequate reporting of reliability by the Counselor's Guide; and (3) other issues regarding reporting of profiles.

Most of the problems of the early forms of ASVAB were subsequently reconciled. Bock and Mislevy (1981) studied the psychometric properties of ASVAB Form 8A using the 1980 American Youth Population and concluded that:

data from responses to the ASVAB are free from major defects such as high levels of guessing or carelessness, inappropriate levels of difficulty, cultural test-question bias, and inconsistencies in test administration procedures. (p. 61)

The ASVAB provides a sound basis for the estimation of population attributes such as means, medians, and percentile points in the youth population as a whole and in subpopulations defined by age, sex, and race/ethnicity. Hunter (1984) has also provided an assessment of military testing:

During the past forty (40) years the military services have conducted extensive validation research on the ASVAB and its predecessors. These studies have established convincing evidence for the validity of the tests which make up the test battery. All together this research provides a data base that shows that the ASVAB as a test battery measures cognitive ability better than most civilian batteries of comparative structure. As a result the validity of the ASVAB is even higher than the validity for batteries which are used in the civilian sector. (p. 1)

Jensen (1985) stated that:

The recently published version—ASVAB Form 14—may be regarded, from a psychometric standpoint, as an exemplar of the state of the art for norm-referenced, group administered paper-and-pencil tests of mental abilities. This is not to say, however, that some of the suggested specific uses of the ASVAB can escape serious criticism. Its potential liabilities concern the manner of its use, not its psychometric features which are generally admirable. Moreover, the total "package" offered by DoD is attractive, impressive, and unmatched by any commercially available test. (p. 32)

Jensen's main criticism addressed the high intercorrelations between ASVAB subtests, the resulting lack of differential validity for the composites formed of ASVAB subtests, and the restricted value of the ASVAB for aiding in individual occupational choices. This criticism, however, applies equally well to commercially available multiple-test batteries. In spite of the high intercorrelations, Jensen contends that utility may be achieved by institutions when they are selecting applicants or students from a large population.

Murphy (1984), in a similar vein as Jensen, conjectured that composite scores are so highly intercorrelated that they rarely differentiate among vocational aptitudes for appropriate placement of recruits into technical schools and, ultimately, service careers. The problem of composite redundancy, due to high composite intercorrelations, is illustrated with ASVAB Form 14 for which the average intercorrelations among the seven composites is .86. Using alternate form reliabilities in the correction formula to correct for attenuation, the average intercorrelation is .92. Such redundancy can be explained largely by the preponderance of composites used in high schools and across military services sharing one or more test. Moreover, while validity studies generally support the contention that ASVAB scores adequately predict success in training schools, little evidence supports the proposition that composite scores are differentially valid for predicting performance in diverse forms of training.

Murphy (1984) contends that the various ASVAB composites developed expressly for a specific training program could be replaced by just one composite without any appreciable decrease in validity. He concludes that there is little evidence that the ASVAB provides more than a measure of general ability. As, in the case of Jensen, this criticism applies equally to all multiple batteries, whether ASVAB or commercial equivalents.

The following references provide additional background information on the ASVAB: Eitelberg, Laurence, Waters, and Perelman (1984), *Test Manual for the Armed Services Vocational Aptitude Battery* (Department of Defense, 1984a), and the *Technical Supplement to the Counselor's Manual for the Armed Services Vocational Aptitude Battery Form-14* (Department of Defense, 1984b).

NOTES

1. ASVAB Form 3 was used by the Air Force and Marine Corps as their operational test; ASVAB Form 4 was designed for the DoD Student Testing Program, but never implemented.

2. ASVAB 14 contains the same tests and is parallel to Forms 8 through 13 of the Production Testing Program.

3. This assignment process is presented here as an idealized concept. Actual classification and assignments are also affected by other factors such as enlistment contracts, guarantees, and aptitude composites.

4. Prior to January 1, 1950, the Navy used the Navy Applicant Qualification Test, Form 3 to evaluate generally similar characteristics of Selective Service inductees or applicants for enlistment (Uhlaner & Bolanovich, 1952).

5. Because of its continued period of use (1950–1984), the score scale was used during the Vietnam War as well as the Korean War.

REFERENCES

ASVAB Working Group. (1980). *History of The Armed Services Vocational Aptitude Battery (ASVAB)* (Tech. Rep. No. SBI-AD-E750 743). Washington, DC: Office of the Secretary of Defense.

Birnbaum, A. (1968). Some latent trait models and their use in inferring an examinee's ability. In F. M. Lord & M. R. Novick (Eds.), *Statistical theories of mental test scores* (Chapters 17–20). Reading, MA: Addison-Wesley.

Bock, R. D., and Mislevy, R. (1981). *Data quality analysis of the Armed Services Vocational Aptitude Battery*. Chicago, IL: National Opinion Research Center.

Bock, R. D., and Moore, G. H. (1984). *Profile of American youth: Demographic influences on ASVAB test performance*. Washington, DC: Office of the Assistant Secretary of Defense.

Booth–Kewley, S., Foley, P., and Swanson, L. (1984). *Predictive validation of the Armed Services Vocational Aptitude Battery (ASVAB) Forms 8, 9, and 10 against performance in 100 Navy schools* (NPRDC Tech. Rep. 85-15). San Diego: Navy Personnel Research and Development Center. (AD-A149 695).

Cronbach, L. J. (1979). The Armed Services Vocational Aptitude Battery—A battery in transition. *Personnel and Guidance Journal, 57*, 232–237.

Department of Defense. (1984a). *Test manual for the Armed Services Vocational Aptitude Battery*. North Chicago, IL: Office of the Assistant Secretary of Defense, United States Military Entrance Processing Command.

Department of Defense. (1984b) *Technical supplement to the counselor's manual for the Armed Services Vocational Aptitude Battery Form–14*. North Chicago, IL: United States Military Entrance Processing Command.

Department of Defense (1986). *A review of the development and implementation of the Armed Services Vocational Aptitude Battery Forms 11, 12, and 13*. Washington, DC: Subcommittee of the Joint-Service Selection and Classification Working Group.

Divgi, D. R., and Horne, G. E. (1986). *Using the high school ASVAB in ninth and tenth grades* (Tech. Rep. No. CNR–119). Alexandria, VA: Center for Naval Analysis.

Earles, J. A., Giuliano, T., Ree, M. J., and Valentine, L. D. Jr. (1983). *The 1980 youth population: An investigation of speeded subtests.* Unpublished manuscript, Brooks AFB, TX: Air Force Human Resources Laboratory, Manpower and Personnel Division.

Eitelberg, M. J., Laurence, J. H., Waters, B. K., and Perelman, L. S. (1984). *Screening for service: Aptitude and education criteria for military entry.* Washington, DC: Office of the Assistant Secretary of Defense.

Hunter, J. E. (1984). *The validity of the ASVAB as a predictor of civilian job performance.* Rockville, MD: Research Applications.

Jensen, A. R. (1985). Test reviews—Armed Services Vocational Aptitude Battery. *Measurement and Evaluation in Counseling and Development, 18,* 32–37.

Kroeker, L. P., and Rafacz, B. A. (1983). *Classification and assignment within PRIDE (CLASP): A recruit assignment model* (NPRDC Tech. Rep. 84–9). San Diego: Navy Personnel Research and Development Center. (AD–A136 907).

Maier, M. H. and Truss, A. R. (1983). *Original scaling of ASVAB Forms 5/6/7: What went wrong* (Tech. Rep. No. CRC 457). Alexandria, VA: Center for Naval Analyses.

_____ . (1985). *Validity of the Armed Services Vocational Aptitude Battery Forms 8, 9, and 10 with applications to Forms 11, 12, 13, and 14* (Tech. Rep. CNR 102). Alexandria, VA: Center for Naval Analyses.

Maier, M. H., and Sims, W. H. (1986). *The ASVAB score scales: 1980 and World War II* (Tech. Rep. No. 116). Alexandria, VA: Center for Naval Analyses.

McLaughlin, D. H., Rossmeissl, P. G., Wise, L. L., Brandt, D. A., and Wang, M. (1984). *Validation of current and alternative Armed Services Vocational Aptitude Battery (ASVAB) area composites* (Tech. Rep. No. 651). Alexandria, VA: U.S. Army Research Institute.

Murphy, K. R. (1984). Armed Services Vocational Aptitude Battery. In D. J. Keiser & R. C. Sweetland (Eds.), *Test critiques.* Kansas City, MO: Test Corporation of America.

Prestwood, J. S., Vale, C. D., Massey, R. H., and Welsh, J. R. (1985). *Armed Services Vocational Aptitude Battery: Development of Forms 11, 12, and 13* (AFHRL TR–85–16). Brooks AFB, TX: Air Force Human Resource Laboratory, Manpower and Personnel Division.

Ree, J. M., Welsh, J. R., Wegner, T. G., and Earles, J. A. (1985). *Armed Services Vocational Aptitude Battery: Equating and Implementation of Forms 11, 12, and 13 in the 1980 Youth Population Metric* (AFHRL–TP–85–21). Brooks AFB, TX: Air Force Human Resource Laboratory, Manpower and Personnel Division.

Sims, W. H., and Maier, M. H. (1983). *The appropriateness for military applications of the ASVAB subtests and score scale in the new 1980 reference population* (Memorandum 83–3102). Alexandria, VA: Center for Naval Analysis.

Stuit, D. B. (1947). *Personnel research and test development in the Bureau of Naval Personnel.* Princeton: Princeton University Press.

Uhlaner, J. E., and Bolanovich, D. J. (November 7, 1952). *Development of the Armed Forces Qualification Test and predecessor Army screening tests,*

1946–1950 (PRS Rep. 976). Washington, DC: Department of the Army, Personnel Research Section.

Wegner, T. G., and Ree, M. J. (1985). *Armed Services Vocational Aptitude Battery: Correcting the speeded subtests for the 1980 youth population* (AFHRL-TR-85-14). Brooks AFB, TX: Air Force Human Resources Laboratory.

Wigdor, A. K., and Green, B. F. (1986). *Assessing the performance of enlisted personnel (Evaluation of a joint-service research project)*. Washington, DC: National Academy Press.

Wilbourn, J. M., Valentine, L. D., Jr., and Ree, M. J. (1984). *Aptitude index validation of the Armed Service Vocational Aptitude Battery (ASVAB) Forms 8, 9, and 10* (AFHRL-TP-84-08). Brooks AFB, TX: Air Force Human Resources Laboratory.

3   The Influence of Paragraph Comprehension Components on Test Validity

Susan Embretson
James Fultz
Nancy Dayl

Paragraph comprehension items are important measures of both verbal ability and reading achievement. Typically a paragraph comprehension item consists of a short passage followed by one or more questions. These item types appear on a variety of important tests, including the Armed Services Vocational Aptitude Battery (ASVAB), the Graduate Record Examination (GRE), and the Gates-MacGinitie Reading Test. However, paragraph comprehension items appear to be complex item types. In the ASVAB, for example, paragraph comprehension subtest has a relatively low reliability as compared to the other subtests (Moreno et al., 1984). Furthermore, item analyses have sometimes revealed that the item discrimination parameters are changing from ASVAB items to the new prototypic items for computerized adaptive testing.

Recent research (Embretson, 1985a; Embretson & Wetzel, 1985) has supported two major stages of cognitive processing on paragraph comprehension items on the ASVAB; text representation and response decision. Embretson and Wetzel (1985) calibrated items on several stimulus complexity factors that they postulated to influence the text representation process and the decision process independently. The stimulus complexity variables for text representation difficulty were obtained from a propositional analysis of the text (Kintsch & van Dijk, 1978) and an analysis of word frequency (Kucera & Francis, 1967). The stimulus complexity variables for decision difficulty were abstracted from cognitive component research on other verbal item types, such as analogies (Sternberg, 1977; Pellegrino & Glaser, 1979; Whitely & Barnes, 1979).

Embretson and Wetzel (1985) used a component latent trait model to calibrate the impact of the stimulus complexity factors on item difficulty. The results indicated that the two stages are empirically independent sources of difficulty in the ASVAB items, as the correlation of processing difficulty on the text and decision processing was near zero. Furthermore, the data also suggested that performance on the stages reflects different ability dimensions.

These findings, taken together, suggest that a paragraph comprehension test may measure either the abilities involved in text representation or the abilities involved in decision processing or some combination, depending on which items are selected for the test. If the relevant stimulus complexity factors are not explicitly controlled in the test specifications, problems in equting alternative test forms can arise. Thus, one form could emphasize primarily text representation while another form could emphasize decision processing. Even worse, computerized adaptive testing, where the items that are administered depend on the person's sequential response pattern, can possibly lead to adaptive tests that measure different combinations of abilities for different persons.

Embretson and Wetzel's (1985) findings also suggest an explanation for the relatively poor item characteristics of paragraph comprehension items on the ASVAB. Since paragraph comprehension items may measure predominantly either text representation ability or decision processing, if the subtest consists of a mixture of extreme items on text and decision, low reliability will be found. Furthermore, if new items are not guided by item specifications about text and decision difficulty, possible drift in the ability that is measured may result. Item discrimination differences would reflect this change.

To investigate the latter, Embretson and Wetzel (1985) compared the cognitive components of ASVAB paragraph comprehension items to a new set of items that had been constructed for the computerized version of the paragraph comprehension subtest (CAT). Although the stimulus complexity factors for the text representation processes were not significantly different, the stimulus complexity for the decision process was significantly more difficult for the new CAT items. These results indicate that the new items are better measures of the abilities that are associated with the decision.

Since items may depend on different processing abilities, it is likely that the validity of an item subset is also influenced by processing components. This is an especially important consideration for a computerized adaptive test, such as planned for the military, since different individuals receive different item subsets. If these subsets vary in cognitive component representation, the validities of the tests may differ. Although the cognitive parameters that were identified in the previous research on paragraph comprehension items can provide the basis for item specifications, it is not clear which processing ability is the most important for test validity. For example, if only items with difficult decision processing and easy test representation are selected (for example, the new CAT prototypic items), will test validity increase or decrease?

The goal of this research is to examine the influence of the cognitive processing characteristics on the construct validity of paragraph comprehension items on the ASVAB. Several subcategories of construct validity will be examined, including convergent, discriminant, predictive, and incremental validity (with respect to the other ASVAB subtests). Items from both the standard ASVAB tests and the computerized adaptive testing bank are studied.

This chapter will be organized into three studies and a general conclusion section. Study 1 will compare empirically several models of cognitive characteristics to select a model for the study. Data from the Embretson (1985a) study on 75 paragraph comprehension items will be reanalyzed. Study 2 compares the construct representation of ASVAB-CAT items to another popular test of paragraph comprehension, the Gates–MacGinitie. That is, the stimulus complexity factors for the text representation component and for the decision component are compared between tests, using the cognitive model selected in Study 1. Study 3 presents results on the impact of cognitive components of paragraph comprehension items on test validity. This section will examine the validity of computer adaptive tests as dependent on the relative representation of cognitive components. The final section presents a summary and conclusions.

STUDY 1: MODELS OF COGNITIVE COMPONENTS
OF PARAGRAPH COMPREHENSION

Studying the cognitive characteristics of paragraph comprehension items depends on having a theoretically and empirically adequate model. Previous research (Embretson & Wetzel, 1985) developed two cognitive component models that gave moderately good prediction item difficulty with the linear logistic latent trait model (Fischer, 1973). The text representation variables were based on a propositional analysis of the text (Kintsch & van Dijk, 1978) using Bovair and Kieras' (1981) and Turner and Green's (1978) system.

The decision process variables were based on a two-stage model of processing multiple alternatives. The first stage is falsification, which is a relatively fast stage in which the examinee attempts to exclude an alternative from further consideration on the basis of a few global features. This stage is followed by confirmation, which requires more extended processing, but is applied only if the alternative is not falsified. Both models included variables for the falsifiability of the distractors, the confirmability of the correct answer, and the difficulty of mapping the alternatives to the text.

Embretson and Wetzel (1985) compared these models to others that had been hypothesized. One alternative model fared particularly well. Anderson (1972) hypothesized that several levels of question-asking about a text were important in measuring comprehension. Embretson and Wetzel (1985) scored Anderson's levels as two variables: (1) inference, a binary variable reflecting whether the alternatives required induction or deduction from the paragraph to falsify or confirm; and (2) paraphrase level, the degree to which the wording of the text corresponds to the wording of the alternative (that is, the variable ranged from verbatim to restructured paraphrases). Good prediction of item difficulty was obtained from these two variables. The high prediction that was achieved from the levels of questioning variables suggests that perhaps these variables should be included in a component model of paragraph comprehension items for the current study.

The purpose of this study is to select a model for examining the influence of the cognitive components of paragraph comprehension items on test validity. Embretson and Wetzel's (1985) data for 75 paragraph comprehension from a large sample of recruits was available for the current study to examine the impact of the levels of questioning variable.

Method

Subjects and Tests

Item response data was available for a large sample of military recruits on 75 ASVAB and CAT paragraph comprehension itesm. Of the items, 29 were from ASVAB and 46 were from CAT.

The design of the military testing data was to administer one ASVAB form and one CAT booklet to each recruit. To obtain information on all the selected items, data from 12 groups was required. The groups were analyzed simultaneously by the linear logistic latent trait model (LLTM). The maximum likelihood item difficulty estimates in LLTM are linked across groups through items that are common to two or more groups.

Cognitive Variables

All cognitive variables were scored for each of 75 items and then used as complexity factors to model item difficulty in LLTM. Since several variables were obtained from a propositional analysis of the text, a brief description will be given here. A complete description of the cognitive variables is given in Embretson and Wetzel (1985).

In propositional analysis (Kintsch & van Dijk, 1978), propositions are composed of concepts, in which the first element is a predicate or relational concept, and the remaining elements are arguments. For example, the sentence "Nan wrote a long essay in order to win the prize" can be propositionalized as follows:

P1. Modify: essay, long

P2. Write: Nan, P1

P3. Win: Nan, prize

P4. In order to: P3, P1

In this sentence, P2 and P3 are predicate propositions. P1 is a modifier proposition, and P4 is a connective proposition. The arguments are the actors and objects, such as Nan and prize. In the current study, the density of propositions was scored by the number of propositions of each type in the paragraph divided by the total number of words.

Model 1 variables for the text consisted argument density, predicate propositional density, connective propositional density, modifier propositional density, and word frequency for all words in the text. Model 1 variables for the decision consisted of falsifiability, confirmability, text mapping for correct answer (that is, log frequency of the number of comparisons to text propositions required to confirm the answer), text mapping for the distractors, and reading grade level (Flesch, 1948) for the distractors.

The Model 2 text variables consisted of argument density, predicate propositional density, connective propositional density, modifier propositional density, word frequency for content words (as opposed to function words, such as articles, prepositions, auxiliary verbs, etc.) and the percent of content words. The Model 2 decision variables consisted of falsifiability, confirmability, text mapping for the correct answer, text mapping for the distractors, percent of text relevant to the alternatives, content word frequency for the correct answer, and content word frequency for the distractors.

Alternative versions of Model 1 and Model 2 were constructed by substituting Anderson's (1972) levels of questioning variables (as described above) for conceptually similar variables in Model 1 and Model 2. First, inference replaced the text mapping variables. The text mapping variables generally measure the difficulty of attaching the alternatives to the text as the number of propositional comparisons required to evaluate the correct answer and distractors. Although text mapping had significant weights in Model 1 and marginally significant weights in Model 2, the direct correlation to item difficulty was weak. The level of questioning categories provide another index of text mapping, since mapping an alternative to the text is more difficult if an inference is required. Thus, the text mapping variables were replaced by variables that indicated if an inference was required to falsify or confirm the alternative.

Second, the paraphrase level variable replaced a similar variable in Model 1 and Model 2, encoding conversion. Encoding converion is a binary variable that is scored if the wording of the text was different than the wording of the alternative. In contrast, the paraphrase level variable contains four levels, ranging from verbatim to restructured paraphrases. In Model 1 and Model 2, encoding conversion was a multiplier that increased the difficulty of falsification and confirmation. In the current study, paraphrase level was substituted for encoding conversion as the multiplier in both models.

Results

Table 3.1 presents the significance and goodness of fit of aternative LLTM cognitive models. The original models are Model 1a and Model 2a, and the alternative versions with the level of questioning variables are Model 1b and Model 2b. The table shows the log likelihood of the data for each model. Further statistics for the models are developed according to Embretson's (1983) incremental fit indexes for the LLTM. A chi square test, χ^2, for significant prediction of item difficulty compares the model to a null model of no prediction and a fit index, Δ, gives the relative amount of information accounted for by the model. The fit index is similar to r^2 in range and magnitude.

For the models containing only the decision variables, Table 3.1 shows that improved fit is indicated for both Model 1 and Model 2 by substituting inference and transformation for text mapping and encoding conversion, respectively, into Model 1b and Model 2b. Further, the chi squares for all models indicate significant prediction, but the chi squares are larger for the levels of questioning variables.

Table 3.1 also shows the significance and goodness of fit for the full models that include both the text and decision variables. Although the fit index shows little change, the levels of questioning variables still have smaller chi square values.

Discussion

The results indicate that when the decision process is considered separately from the text representation variables, the levels of questioning variables improve fit in both Model 1 and Model 2. However, in the context of the full model, including the text, the fit improves only slightly, which indicates a correlation between some sources of text difficulty and the level of questioning variables.

Table 3.1
Significance and Goodness of Fit from Alternative Cognitive Models

Model	lnL	X^2	df	fit Δ
Decision Variables				
Model 1a	−26624.53	2045.54	5	.28
Model 1b				
	−26570.57	2153.46	5	.30
Model 2a	−26727.87	1838.36	7	.25
Model 2b				
	−26618.67	2057.26	7	.28
Full Models				
Model 1a	−26330.33	2633.94	10	.36
Model 1b				
	−26314.23	2666.14	10	.37
Model 2a	−26317.07	2660.46	13	.37
Model 2b				
	−26305.21	2684.13	13	.37

Several considerations guided selection of a model from Table 3.1 for the validity study. For empirical adequacy, it was desirable not only to have the highest level of prediction possible but also to minimize the number of independent variables in a model. Model 2b with the levels of questioning variables had the highest chi square (as compared to a null model), which indicates the best fit. Model 1b with the levels of questioning variables, on the other hand, had nearly as high chi square from only ten variables. For theoretical adequacy, it was desirable to select the model in which the independent variables represented relatively unconfounded sources of stimulus complexity. In both Model 1a and Model 1b, word frequency as a measure of vocabulary level is confounded with sentence structure because the (high) frequency of function, as well as the content words, is included in the average total word frequency. A difficult text that contains infrequent content words, which convey essential meaning, can still have a high average word frequency if it contains several high frequency function words, such as auxiliary verbs or articles. Furthermore, in both Model 1a and Model 1b, reading grade level (although a good predictor) is also a confounded index, since sentence length is combined with word length in the index. Both Model 2a and Model 2b, in contrast, do not contain these confounded measures. Thus, Model 2b was selected as the most theoretically and empirically adequate model.

Since Model 2b will be used to examine the impact of text representation and decision processing on validity, the variables in the model will be discussed here. Table 3.2 shows the LLTM weights for each stimulus complexity factor in item difficulty for Model 2b. These weights are similar to unstandardized coefficients in multiple regression. The dependent variable is the exponential item difficulty scale of the Rasch model. Also presented on Table 3.2 are the significances associated with each weight and the simple correlation for each stimulus complexity factor with item difficulty. For text representation, it can be seen that argument density and percent of content words have positive relationships to item difficulty, while the density for connective propositions has a negative relationship. Thus, difficult items have texts that have many different arguments, a high proportion of content words and relatively few connective propositions. A more detailed discussion of these effects is available in Embretson and Wetzel (1985).

For the decision process variables, it can be seen that both falsifiability and confirmability have significant negative relationships with item difficulty. Since these variables contain paraphrase level as a multiplier, the negative relationship to item difficulty implies that paraphrasing makes falsification and confirmation more difficult. Also, word frequency for the distractors (but not the correct answer) has a significant negative relationship to item difficulty, indicating that difficult items have relatively infrequent words in their distractors but not the correct answer. Last, inferences required to confirm the correct answer has a significant positive relationship to item difficulty. Inferences required to falsify the distractors, although showing a positive simple correlation, have a significant negative weight. The change of sign suggests that its role in the model is a suppressor variable, through its high correlation with the other inference variable.

STUDY 2: COGNITIVE CHARACTERISTICS OF PARAGRAPH COMPREHENSION ITEMS

Cognitive component analysis of items provides a new way to examine construct validity. Research on many item types (for example, Sternberg, 1985; Pellegrino & Glazer, 1979) have shed new light on the nature of ability by explicating the processing components that underlie performance. Embretson (1983) notes that the conceptualization of construct validity needs to be broadened to include two separate type of research, construct representation and nomothetic span. Construct representation research

Table 3.2
Linear Logistic Weights and Standard Errors for Model 2

Variable	r	η	SE_{η}	t
Text				
Word Frequency, Content	.014	.048	.100	.480
Argument Density	.161	1.478	.448	3.293**
Connective Density	-.205	-5.575	.508	-10.954**
Predicate Density	-.020	-.775	.547	-1.414
Modifier Density	.174	.693	.561	1.234
Percent Content	.272	.821	.250	3.271**
Decision				
Inference, Correct	.356	.593	.166	3.572**
Inference, Distractor	.112	-.327	.165	-1.977*
Percent Relevant Text	.175	.057	.210	.271
Falsification	-.186	-2.319	.692	-3.351**
Confirmation	-.405	-3.229	.406	-7.949**
Word Frequency, Correct	-.121	.147	.149	.980
Word Frequency, Distractor	-.274	-.390	.156	-2.495**

* $p < .05$

** $p < .01$

seeks to explicate the processing components, strategies, and knowledge structures that are involved in item solving. Task decomposition, such as typical of cognitive component analysis, is the major tool for construct representation research. Nomothetic span research, in contrast, seeks to examine the utility of the test for measuring individual differences. Relationships of the test score to other tests, criterion measures, and demographics explicate nomothetic span. Especially important in nomothetic span research is to establish how the relative representation of the underlying cognitive constructs influences the pattern, span, and magnitude of relationships of the test score.

The explication of construct representation by cognitive components is also important for test design. Embretson (1985b) shows two ways in which cognitive components can be used in test design. First, if a component latent trait model is applied (Embretson, 1983), the parameters reflect the impact of cognitive components on items. These parameters can be used to select items with specified sources of cognitive complexity. Second, cognitive characteristics can also be used to guide test design by providing a basis of item specifications. Limits or patterns of cognitive complexity can be specified so that the desired ability is measured.

One interesting implication of this view of test design is that the same type may measure quite different abilities, depending on the explicit (or implicit) specifications that guided item development. For paragraph comprehension items, it is possible to select items subsets that measure predominantly the paragraph representation process by selecting items that are extremely easy on the decision process. Thus, the decision will have little effect on item responses. Conversely, it is also possible to develop a test that measures predominantly the decision process by selecting items that are extremely easy on text representation.

Construct representation may be further explicated by comparing different tests on cognitive characteristics. For example, the ASVAB and CAT items can be compared on cognitive components to another test that is an important measure of paragraph comprehension.

As noted above, the item characteristics of the ASVAB paragraph comprehension items are problematic. A more systemaic basis of test design would be highly desirable. The purpose of the first study is to contribute to test design be elaborating the construct representation of the paragraph comprehension items. Descriptive statistics on the means, standard deviations, and intercorrelations of cognitive components will be presented and compared to another test for measuring paragraph comprehension test, the Gates-MacGinitie Test.

Method

Tests

The military paragraph comprehension item bank for this study consisted of 63 items. About half the items had been included in Study 1. For all items, the text could fit on a CRT display and only one question was asked for each paragraph. The Gates-McGinitie Test consisted of 43 items. Often the text consisted of more than one paragraph and several questions were asked for each paragraph.

Cognitive Variables

Two graduate students in psychology independently propositionalized the text and alternatives of the 63 military items and the 43 Gates-MacGinitie items and scored the decision process variables.

The raters then scored their protocols for number of modifier, predicate, connective, and total propositions as well as the number of unique arguments that occur in the propositions. The decision process variables that required raters included: (1) encoding conversion for arguments, the requirement to convert the wording of the arguments in the alternatives to the arguments of the text; (2) encoding conversion for predicates, the requirement to convert the wording of the predicates in the alternatives; (3) confirmation, scored positively if the text confirms the correct answer; (4) the number of alternatives that are falsified by the test; and (5) the level of questioning of the variable, according to a six-category classification from Anderson (1972).

Results

Reliability

High reliability was obtained for the variables scored by raters. Reliability for the four propositional analysis variables, argument density, modifier density, predicate density, and connective density ranged from .92 to .99 for the ASVAB-CAT and from .96 to .99 for the Gates-MacGinitie. Encoding conversion, confirmation, and number of falsifiable distractors showed reliabilities of .89, 1.00, and .93 respectively, for the ASVAB-CAT, and reliabilities of .83, 1.00, and .96 respectively, for the Gates-MacGinitie. A cross-tabulation of the levels of questioning categories, by rater, found rater agreement was 90 percent for the ASVAB-CAT and 80 percent for the Gates-MacGinitie.

Construct Representation

Table 3.3 presents the means and standard deviations of the text and decision variables from Model 2b for the ASVAB-CAT and Gates-MacGinitie items. A multivariate analysis of variance indicated that both text and decision sets differ significantly between the tests (p's < .01). The univariate analyses of variance are shown on Table 3.3.

Table 3.3 shows that several text and decision variables differ significantly between the tests. For the text variables, the univariate analyses of variance indicate that the Gates-MacGinitie is characterized by significantly higher predicate density, higher percentage of content words, and a lower word frequency. For the decision variables, the Gates-MacGinitie has a significantly higher falsifiability for the distractors, higher confirmability of the correct answer, lower percentage of relevant text, and lower word frequencies for both the correct answer and distractors.

The intercorrelations of the text and decision variables present additional information about the implicit design of the tests.

Table 3.4 shows the intercorrelations of text variables for the ASVAB-CAT. Content word frequency has small negative correlations with several

Table 3.3
MANOVA of Text and Decision Variables from Model 2

Variable	GATES		CAT		
	$\overline{X}$	S	$\overline{X}$	S	F
Text					7.55**
Modifier Density	.17	.10	.18	.06	.59
Predicate Density	.17	.11	.14	.03	5.53*
Connective Density	.15	.08	.15	.04	.09
Argument Density	.38	.19	.42	.08	1.22
Word Frequency	583.90	210.00	.38	46.31	4.25**
Percent Content					
Words	.74	.08	.35	.32	24.16**
Decision					
Inference,					
Distractors	.22	.44	.42	.61	3.36+
Inference,					
Correct	.21	.41	.32	.47	1.49
Falsification	.08	.04	.06	.03	21.09**
Confirmation	.28	.11	.22	.03	17.54**
Percent Relevant					
Text	.17	.12	.52	.30	53.62**
Word Frequency,					
Correct	405.49	820.44	974.01	803.22	14.72**
Word Frequency,					
Distractor	389.08	582.96	1001.56	700.25	26.73**

variables, while the largest correlation is between modifier density and argument density. Table 3.5 shows the Model 2b decision variables. The highest correlation is between the two inference variables, followed by a high negative correlation of confirmation with inference for the correct answer and then a high correlation of word frequency for correct answer with word

Table 3.4
Intercorrelations of Text Cognitive Variables from Model 2 for CAT Items

Text	(1)	(2)	(3)	(4)	(5)	(6)	(7)
1. Word Frequency, Content	1.00						
2. Argument Density	-.36	1.00					
3. Connective Density	-.09	.30	1.00				
4. Predicate Density	.12	-.23	-.29	1.00			
5. Modifier Density	-.39	.58	-.29	-.14	1.00		
6. Number of Words	-.08	-.16	.01	-.04	.00	1.00	
7. Percent Content	-.35	.31	-.06	.10	.28	-.30	1.00

frequency for distractors. Percentage of relevant text had moderate positive correlations with the two inference variables.

The correlations between the text and decision variables for the Gates–MacGinitie items are presented on Table 3.6 and Table 3.7. It can be seen on these tables that, for the text, the propositional density variables have high positive correlations on the Gates–MacGinitie and that percent of content words is negatively correlated with content word frequency. For the decision, the inference variables are highly intercorrelated as are the text mapping variables. Falsification and confirmation are also highly correlated.

Table 3.5
Intercorrelations of Decision Variables from Model 2 for CAT Items

Decision	1	2	3	4	5	6	7
1. Inference, Correct	1.00						
2. Inference, Distractor	.87	1.00					
3. Percent Relevant Text	.47	.43	1.00				
4. Falsification	-.43	-.38	-.18	1.00			
5. Confirmation	-.74	-.39	-.34	.31	1.00		
6. Word Frequency, Correct	.00	-.04	-.09	-.09	-.04	1.00	
7. Word Frequency, Distractor	-.01	-.02	.04	-.13	-.09	.67	1.00

Table 3.6
Intercorrelations of Text Cognitive Variables from Model 2 for Gates Items

Text	(1)	(2)	(3)	(4)	(5)	(6)
1. Word Frequency, Content	1.000					
2. Argument Density	.181	1.000				
3. Connective Density	.019	.914	1.000			
4. Predicate Density	.234	.948	.823	1.000		
5. Modifier Density	.114	.929	.808	.910	1.000	
6. Percent Content	-.770	-.059	.008	.033	-.023	1.000

Table 3.8 presents descriptive statistics on the overall difficulty of text and decision on the 63 CAT-ASVAB items as compared to Embretson and Wetzel's (1985) 73 items. For both data sets, Model 2b was used as stimulus complexity factors to predict the stimulus complexity of text and decision. For Embretson and Wetzel (1985) items, which were used to derive the LLTM weights, the mean item difficulty is set at .0000 on the logit scale of item difficulty, as is standard in the LLTM model. However, for the current data set, the LLTM weights are applied to the new 63 items, as in a cross-validation. It can be seen on Table 3.8 that both text and decision difficulty is somewhat lower in the 63 CAT-ASVAB items, and that text and decision difficulty now have a slight negative correlation, rather than a positive correlation. Further, the standard deviations are somewhat smaller.

Table 3.7
Intercorrelations of Decision Variables from Model 2 for Gates Items

Decision	(1)	(2)	(3)	(4)	(5)	(6)	(7
1. Inference, Correct	1.000						
2. Inference, Distractor	.913	1.000					
3. Percent Relevant Text	-.006	.017	1.000				
4. Falsification	-.354	-.360	-.128	1.000			
5. Confirmation	-.382.	-.373	-.027	.552	1.000		
6. Word Frequency, Correct	-.080	-.079	.036	.469	.101	1.000	
7. Word Frequency, Distractor	.146	.046	-.141	.281	.065		1.00(

Table 3.8
Means and Standard Deviations for Text and Decision Difficulty

	CAT Items (N=63)			CAT Prototype (N=75)		
	$\overline{X}$	sd	r	$\overline{X}$	sd	r
Text	-.120	.384		.000	.443	
Decision	-.095	.284	-.173	-.005	.310	.233

Discussion

The results indicate that the two tests have different sources of cognitive complexity. The Gates–MacGinitie Test has lower word frequencies, which should make vocabulary a more important factor, but it has higher density of predicates and content words, which should make the text easier to parse for meaning than ASVAB-CAT test. In the context of the cognitive model of item difficulty, the higher percentage of content words is associated with higher item difficulties.

On the other hand, the CAT is particularly difficult on decision process variables related to linking the alternatives to the text, but easier on variables-related vocabulary (that is, word frequency). That is, falsifiability of the distractors is lower, the confirmability of the correct answer is lower, and the percentage of relevant text is higher. Also, a marginally significant larger mean for inference on the distractors was found for the CAT items. A cognitive model, Model 2b, indicated that relationships in this direction led to higher item difficulty. However, the word frequencies of both the correct answer and distractors were lower for the Gates–MacGinitie. Low word frequency for distractors is associated with high item difficulty.

To summarize, the results generally suggest that the Gates–MacGinitie Test is more focused on semantic information that underlies reading comprehension while the CAT is more focused on verbal reasoning and, possibly, parsing meaning with complex syntactic information. The text for the CAT items is characterized by a large number of function words (articles, prepositions, etc.) and a relatively high word frequency. The high word frequency indicates easier semantic accessibility of the words in the text, but the sentences contain relatively more syntactic information that is not essential to meaning, as indicated by the larger percentages of function words in the text.

However, the major source of difficulty for the CAT items appears to be the decision process. Although the CAT items have higher word frequencies for the alternatives, as in the text, the distractors are difficult to falsify by the text and the correct answer is difficult to confirm. That is, the number of falsifiable distractors is less and the paraphrasing level is higher.

Further, the implicit design principles vary somewhat between the two tests. That is, the correlations of the cognitive variables among the items differ between the tests. Thus, the specific combination of sources of cognitive complexity is not represented equally between the tests. For example, in the Gates–MacGinitie, the word frequency of content words has a high negative correlation with the percentage of content words. This correlation is much smaller for the CAT items.

Last, a comparison of the CAT-ASVAB with the Embretson and Wetzel (1985) items, which was used to develop the model in Study 1, indicated that CAT-ASVAB items were somewhat easier and less variable on both text and decision difficulty. Further, text and decision difficulty now have a slightly negative correlation, rather than a positive correlation.

STUDY 3: COGNITIVE CHARACTERISTICS AND THE VALIDITY OF THE COMPUTERIZED ADAPTIVE PARAGRAPH COMPREHENSION TEST

Study 1 and Study 2 indicated that the construct representation of the CAT paragraph comprehension item bank is influenced by both the text and the decision complexity. To establish impact on construct representation, the complexity of the text and the decision components must be linked to item responses. The results from the linear logistic latent trait model (reported in Study 1 and in Embretson & Wetzel, 1985) indicated that cognitive complexity of both text and decision variables are reasonably good predictors of the item difficulty parameter.

Further, however, the analyses of the CAT-ASVAB items indicated that the items in the bank varied in the source of cognitive complexity. The means and standard deviations for the text and decision difficulty were about the same magnitude, indicating that the text and the decision complexity had about the same impact on item difficulty. Further, text complexity and decision complexity had a very small negative correlation, indicating that text and decision are relatively independent sources of complexity. However, it should be noted that the items did not vary as much as Embretson and Wetzel's items, and that text and decision had a small positive correlation in their items.

Thus, these data indicate that the CAT-ASVAB items vary in construct representation. Since previous research had indicated that text and decision processes are associated with different abilities, the varying construct representation of the items may, in turn, influence empirical validity.

In Embretson's (1983) conceptualization of construct validity, construct representation influences nomethetic span, which is a summary of empirical validity. That is, nomothetic span is the utility of a test for measuring individual differences, which is indicated by the magnitude, pattern, and span of correlations of the test scores with other measures of individual differences (that is, other tests, criterion measures, demographics, etc.). For a test with variable item content, the impact of varying construct representation in the item bank may have unintended effects on nomothetic span. In computerized adaptive tests, such as for the CAT-ASVAB, each individual receives a small number of items from the item bank to maximize information about ability. The individual items sets often do not even overlap under computerized adaptive testing. Equating the ability scores between individuals relies on item response theory, which entails strong assumptions about the equivalency of the items.

The results from the CAT paragraph comprehension items show that it is quite possible to select items in which difficulty arises predominantly from text representation complexity or, in contrast, from decision complexity. The nomothetic span of the test scores will depend on the particular combination of items that are administered if the text and decision abilities show different patterns of correlations with other measures.

The possible impact of the cognitive characteristics of an item set on validity presents a difficulty for computerized adaptive testing. The equating of item sets between individuals will not be totally effective if an additional property of items, such as the source of cognitive complexity influences nomethetic span.

The current study examines the impact of cognitive characteristics on validity for items sets that are administered from a computerized adaptive test. Perhaps the most important aspect of validity for the military Paragraph Comprehension test is the prediction of grades in military specialty training. This will be a primary focus of the current study. However, the impact of the cognitive charactersitics of item sets on incremental validity, discriminant validity, and convergent validity will also be examined.

Method

Subjects

The subjects were 3,940 military recruits who had taken a computerized adaptive form of the Paragraph Comprehension test to measure ability (CAT-PC ability). These tests consisted of only ten items which were selected adaptively for each recruit to provide maximal information about ability (Booth-Kewley et al., 1985). The subjects had been enrolled in one of 13 military training schools. The preenlistment ASVAB scores were also available for all recruits, while military specialty training grades were available for 2,206 recruits in six schools.

Independent Variables

The potential impact for the cognitive characteristics of adaptive tests depends on the extent to which the administered item sets actually vary in item content between recruits. The variability in the item sets that were administered by the computerized adaptive testing procedure was examined prior to scoring the tests for cognitive characteristics. If the actual items that were administered varied little between recruits, the potential impact of the cognitive characteristics of items would be minimal. A test was scored as unique if the items that were administered did not completely overlap with another test. It was found that 2,369 unique tests were administered to the 3,940 recruits. Of the unique tests, 1,794 tests were administered to only one recruit each, while 278 tests were administered to just two recruits each. Of the remaining 297 tests, the most recruits receiving any single test was 30. Thus, the data clearly indicate substantial variability in which items are received by the recruits in the current study.

Each recruit's computerized adaptive test was scored for the average difficulty of the items on text and decision. Scores for the 63 CAT-ASVAB items in the item bank on stimulus complexity factors for the text and decision were already available from Study 2, using linear logistic latent trait weights for Model 2b. Thus, for each recruit, the average difficulty of the ten items that were administered on both text and decision was calculated.

Test Scores

Table 3.9 shows the means and standard deviations for the grades and the CAT-PC ability scores from the computerized adaptive test for six military schools. It can be seen that the mean ability scores and their correlations with grades varies between the schools. Furthermore, however, mean grades also vary, presumably reflecting varying grading practices between schools.

Table 3.9
Means and Standard Deviations for CAT Ability and Final Grade for Six Schools

School Code	Grade $\bar{X}$	SD	CAT Ability $\bar{X}$	SD	r
1	87.57	6.00	.45	.58	.23
2	82.73	5.26	.55	.62	.50
3	77.73	6.56	.49	.58	.47
4	87.39	6.21	.56	.67	.49
5	90.21	2.76	.30	.63	.30
6	86.03	5.72	.01	.63	.25

Design

For each recruit, the data consisted of military school grade (the primary dependent variable), two scores for the item set to measure complexity on the text and decision components, the paragraph comprehension ability score from the computerized adaptive test (CAT-PC ability), and the preenlistment ASVAB scores.

Results

Test Complexity

The data indicate that the text and decision components have similar means ($-.128$ and $-.113$, respectively) and standard deviations ($.094$ and $.107$, respectively). The standard deviations are quite small, indicating that the adaptive tests, on the average, did not differ sharply in source of cognitive complexity.

The cognitive complexity scores for the item sets were used to classify each recruit's item set into one of four categories for the pattern of cognitive complexity: (1) Both Low, both the text and decision complexity were below the test means; (2) Decision High, the decision complexity is greater than or equal to the mean while the text complexity is below the means; (3) Text High, the text complexity is greater than or equal to the mean and the decision complexity is below the mean; and (4) Both High, both text and decision are greater than or equal to their respective means.

Table 3.10 shows the means, standard deviations, and sample sizes for the CAT-PC abilities for the four categories of cognitive pattern. It can be seen that ability is lower when both the text and decision complexity are low

Table 3.10
Means, Standard Deviations and N for Cognitive Pattern Groups

Group	Ability		
	$\overline{X}$	SD	N
Both Low	$-.01$	.42	294
High Decision	.77	.44	389
High Text	.01	.52	352
Both High	.86	.64	270

and that ability is relatively high when both the text and decision complexity are high. This is to be expected from an adaptive test, in which the items are selected to be close to the person's ability level. However, Table 3.10 shows that individuals who receive items that are complex only on the text component have abilities that are about the same as individuals who receive items that are easy on both components.

Predictive Validity

Figure 3.1 shows the mean grades by school and cognitive pattern. In general, it can be seen that grades are highest when the tests received were complex on both components and, conversely, grades are lower when neither component was complex. Furthermore, grades are lower when the test was complex only on the text component and grades are higher when the item set was complex only on the decision component. Figure 3.1 mirrors the effects of ability differences between the four cognitive patterns; thus, recruits with higher abilities are receiving higher grades.

A comparison of the four cognitive patterns on grades must adjust scores for the differing ability levels between groups and beween schools. A 4×6 analysis of covariance examined the impact of the cognitive categories on the prediction of military school grade. The independent variables were Cognitive Pattern (4) and School (6). The covariate was CAT-PC ability and the dependent variable was military school grade. This analysis of covariance design adjusts the military school grade for the ability score, using the pooled-within cell regression of grade on ability. Thus, the adjusted dependent variable is a residual, the school grade that is not predicted by the ability score. The analysis determines if the independent variables, Cognitive Pattern and School, contribute to the prediction of school grade beyond the CAT-PC ability score.

Table 3.11 presents results on the uniformity of regression test for the analysis of covariance. It can be seen that ability interacts significantly with both School and Cognitive Pattern, indicating that the relationship of grade to ability is significantly different.

Since the uniformity of regression assumption was not met, Figure 3.2 presents the regression of grade z-scores, within schools, on ability. It can be seen that grades regress the most sharply on abilities when the tests are difficult on the decision. The Text High and Both Low patterns provide less predictive information about grades, and regress to lower means.

Multiple regression analysis was also performed on the pooled-within school correlations to further examine predictive validity. In these analyses, a stepwise regression is performed to examine the contributions of the CAT-PC ability score, text complexity score, decision complexity score, and the various interactions to the prediction of military school grade. The interactions were obtained as the simple products of the text and decision scores with each other and with ability. To compare these analyses to the analyses of covariance presented above, two major differences must be noted: (1) the

Figure 3.1
Mean Grade for Paragraph Model 2

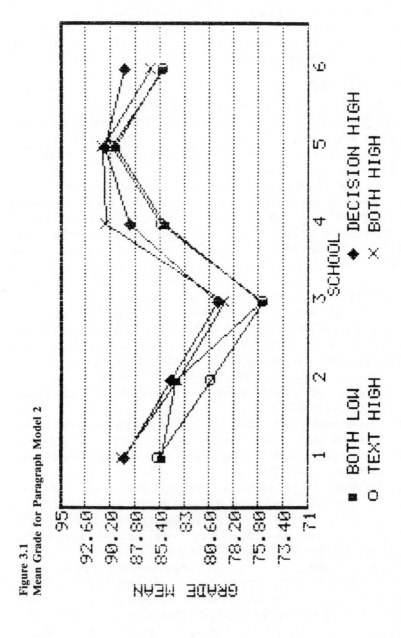

Table 3.11
Analysis of Covariance for Grade by Cognitive Pattern (Model 2), School and CAT Ability

Source	df	SS	MS	F
			Sequential	
Ability	1	3839.83	3839.83	137.40***
School	5	33454.94	6690.98	239.42***
Cognitive Pattern	3	168.45	56.15	2.00
School By Cognitive Pattern	15	750.65	50.04	1.79*
Ability By School	5	748.29	149.65	5.35***
Ability by Cognitive Pattern	3	368.64	122.88	4.39**
Ability By School By Cognitive Pattern	15	273.26	18.21	.65
Residual	1257	3529.69	28.07	

effects of cognitive complexity is measured by continuous variables (text and decision complexity and their interaction) rather than by categories of cognitive complexity, and (2) the effect of the cognitive variables is examined simultaneously with ability, rather than sequentially.

Table 3.12 shows the regression analyses on the pooled-within school data for Model 2. The addition of text difficulty and decision difficulty to ability significantly increases prediction ($F = 3.37$, $p. < .05$) over ability alone. In this set, text difficulty has a significant negative weight. That is, the predicted grade should be lowered if the recruit has been administered an item set that has high difficulty on the text. The addition of the text and decision interaction has a marginally significant effect (Set III) while the addition of the interactions of text and decision with ability is significant ($F = 6.95$, $p < .01$). It should be noted that the absolute increment in the multiple correlation is quite small in all analyses.

Discriminant, Convergent, and Incremental Validity

The final set of analyses use structural equation models to examine further effects on validity. Cognitive Pattern, as defined above, categorized recruits into four groups. The covariance matrixes between CAT ability score, grade (z-score within school), and the ten scores from the preenlistment ASVAB were compared across the cognitive pattern groups in the structural equation analyses.

Discriminant validity was studied from the covariance of the CAT-PC ability with the other ASVAB tests, which measure different constructs. Convergent validity was studied by the covariances of the CAT-PC ability

Figure 3.2
Regression of Grade on Ability, Model 2

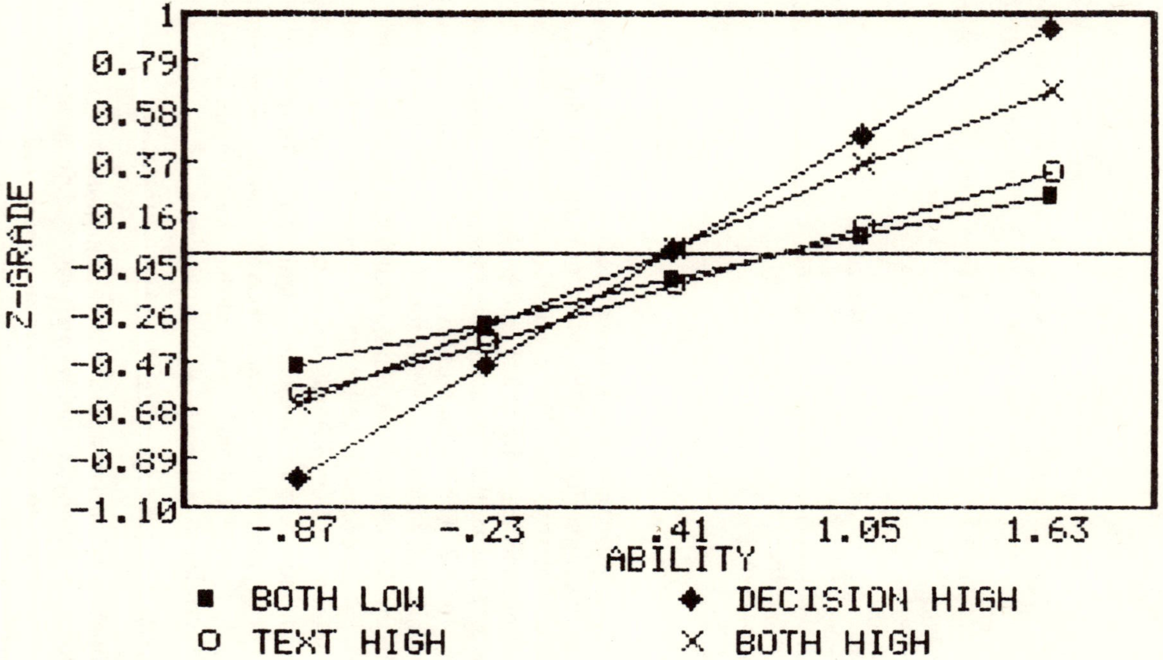

Table 3.12

Regression of Grade on Ability and Cognitive Variables from Pooled-Within School Data (Model 2)

	r	β	t	R	F	F
Set I						
Cat Ability						
Estimate	.38	.38	15.08	.38	227.36***	227.36***
Set II						
Cat Ability						
Estimate	.38	.36	8.84**	.39	77.88***	3.37*
Text Difficulty	-.10	-.06	- 2.17*			
Decision Difficulty	.31	.03	.72			
Set III						
Cat Ability						
Estimate	.38	.37	9.05***	.39	59.41***	3.57+
Text Difficulty	-.10	-.11	- 2.85**			
Decision Difficulty	.31	-.05	- .87			
Text x Decision	-.17	-.10	- 1.89+			
Set IV						
Cat Ability						
Estimate	.38	.38	5.88***	.40	42.38***	6.95**
Text Difficulty	-.10	-.04	- .59			
Decision Difficulty	.31	-.04	- .57			
Ability x Text	-.33	-.09	- 1.28			
Ability x Decision	-.03	.09	3.41***			
Text x Decision	-.17	.00	- .02			

59

with the ASVAB Paragraph Comprehension test, a measure of the same construct. Incremental validity was studied by the significance of the regression of grade on CAT-PC ability, when the nine ASVAB measures of other constructs are also included as predictors.

Table 3.13 shows the goodness of fit for various structural equation models when the cognitive pattern group is based on Model 1 scores for text and decision. The first model, General 1, postulates that both the discriminant validity and the convergent validity are equal between groups and that the incremental validity of the CAT paragraph comprehension test is zero. Model General 1 was specified in the following way: (1) equal discriminant validity is specified by constraining the covariances between the CAT Paragraph Comprehension ability with the nine ASVAB scores across cognitive pattern groups; (2) equal convergent validity is specified by constraining the covariance between the CAT Paragraph Comprehension test and the ASVAB Paragraph Comprehension test across cognitive pattern groups; and (3) zero incremental validity is specified as a regression weight that is fixed to zero for the CAT Paragraph Comprehension ability when the other nine ASVAB scores are included. Table 3.13 shows that this model departed significantly from the data.

The next model, General 2, is identical to General 1, except that incremental validity is not equal to zero. Significant improvement in fit was obtained, indicating that the CAT Paragraph Comprehension ability score significantly increases prediction when added to the nine other subtests on the ASVAB.

The next model postulates that incremental validity is specific to the cognitive pattern group. Although the model still does not fit overall, the fit is not significantly better as compared to the preceding model. The next

Table 3.13
Structural Equation Models for Impact of Cognitive Pattern (Model 2) on Validity

Model	df	x^2	Fit Index	Δdf	Δx^2
General[1] (CAT Incremental Validity = 0)	236	372.05***	.945	--	--
General[2] (CAT Incremenatal Validity $\neq$ 0)	235	357.09***	.948	1	14.96***
Pattern Specific Incremental Validity	232	355.02***	.948	3	2.07
Pattern Specific Discriminant Validity	205	310.25***	.950	30[1]	46.84*
Pattern Specific Validities	202	308.32***	.950	3	1.93

model specifies that both discriminant and convergent validity depend on the cognitive pattern group. In this model, the incremental validities are constrained across groups. Significantly better fit is achieved, as compared to the model General 2. The last model postulates that all validities depend on the cognitive pattern. The model is not significantly better than the preceding model, which specified only that discriminant and convergent validities differed between groups.

Table 3.14 shows the correlations between the CAT Paragraph Comprehension ability and the ASVAB scores for the four cognitive groups. Although convergent validities with the ASVAB Paragraph Comprehension subtest did not differ between cognitive patterns, it can be seen that the highest convergent validity is achieved when the item set has the decision component as the primary source of difficulty. In contrast, CAT items are less correlated with ASVAB when the text difficulty is high.

The discriminant validities did vary significantly between cognitive patterns, and generally higher correlations with ASVAB are found when the decision component is difficult. Table 3.14 shows one exception: CAT-PC ability is most highly correlated with Numerical Operations, when the text is the primary source of difficulty.

Table 3.14
Discriminant and Convergent Validities for Cognitive Pattern (Model 2)

	Both Low	High Text	High Decision	Both High
General Science	.26**	.27**	.40**	.43**
Arithmetic Reasoning	.18**	.24**	.26**	.13**
Word Knowledge	.32**	.36**	.46**	.57**
Numerical Operations	.20**	.37**	.29**	.38**
Coding Speed	.11*	.07	.00	.00
Auto and Shop Information	.07	.12*	-.01	.00
Mathematics Knowledge	.08	.09	.30**	.24**
Mechanical Comprehension	.23**	.26**	.30**	.20**
Electronics Information	.14**	.16**	.29**	.24**
Paragraph Comprehension	.14**	.17**	.39**	.30**

Discussion

The results clearly indicate that the actual items that were received in the current study with the CAT-ASVAB vary substantially between recruits. The computerized adaptive algorithm employed in the current study led to the 3,940 recruits receiving 2,369 tests that did not completely overlap in content.

The current study is apparently the first one to examine the impact of adaptive testing on the cognitive characteristics of the Paragraph Comprehension test on the ASVAB. It was found that the Paragraph Comprehension tests do vary in their representation of text and decision processes, but that this variability is not as extreme as could be obtained from the item bank. Furthermore, the complexity of the text was not adapted to the CAT-PC ability as compared to the complexity of the decision. That is, the mean CAT-PC ability differences between the cognitive patterns indicated that high ability recruits were more likely to receive complex decision processing in their tests than complex text processing. Thus, the CAT-PC scores in the study are more likely to reflect decision processing ability.

The results also indicated that the varying representation of text and decision processes in the adaptive tests also had an impact on validity. Significant differential validities were found for cognitive pattern on both predictive and discriminant validity, but not for convergent and incremental validity.

Predictive validity was examined by two separate analyses, the analysis of covariance and multiple regression. All analyses indicated that significant information for predicting military school grades is given by the cognitive characteristics of the test upon which the ability score is based. However, the effect was not very large.

The analysis of covariance indicated that the cognitive pattern of the test had differential validities in predicting training school. An examination of the separate regression of grade on ability showed that tests that were not difficult on the decision provided less predictive information about grades. Interestingly, items that were difficult on the text were no better in predicting grades than tests that were easy on both text and decision.

The multiple regression analyses show that information about the source of cognitive complexity in the test on which the ability score is based significantly increases the prediction of school grade. That is, significant effects for the text and decision complexity were found. However, the regression analyses reveal that although the impact of cognitive complexity is significant, the magnitude of the impact is generally quite small.

The most compelling conclusion to be drawn from these analyses is that text complexity, as manipulated in the current CAT-ASVAB, is not as central to test validity as the decision processing. Tests that were difficult on the text were associated with lower ability scores and lower predictive validities.

Discriminant, convergent, and incremental validity were studied by structural equation models across groups of recruits who received item sets with different cognitive patterns. It was found that the discriminant validity depended on the cognitive pattern, while convergent and incremental validity did not. For discriminant validity, the cognitive pattern of the test led to different patterns of correlations of CAT-PC ability with the ASVAB subtests. Generally higher correlations, and hence lower discriminant validity, were found for tests that were complex on the decision component. Although incremental validity was found generally for the CAT-PC score in predicting school grade when added to the other ASVAB subtests, the levels of incremental validity did not vary significantly between the cognitive pattern groups.

These results indicate that not only does the source of the cognitive complexity vary across item sets that are administered under computerized adaptive testing, but that the source of cognitive complexity influences the nomothetic span of the test. Although the nature of the impact is not large, and needs to be clarified by further analyses, the results suggest that further safeguards need to be employed in computerized adaptive testing to assure that scores are truly equated between tests that vary in content.

GENERAL CONCLUSION AND SUMMARY

This chapter has examined the impact of the cognitive characteristics of paragraph comprehension items on the validity of ability scores derived from computerized adaptive testing. Under adaptive testing, the item content of the tests varies between examinees since the items are tailored to be optimal for their ability level. With variable item content, the cognitive characteristics of the items sets may also vary between examinees. This could be a serious problem for test validity in the military Paragraph Comprehension test since previous research had found that different abilities are associated with the two major cognitive components of item solving. These components are text representation and decision.

The current study finds that the cognitive characteristics of the item set influence the predictive validity of an adaptive Paragraph Comprehension test. A comparison between item sets with different primary sources of cognitive complexity indicated that a complex test representation component leads to lower predictive validity than a complex decision component. Although the effect was small, all analyses showed that the cognitive characteristics of the item set had significant impact on the prediction of military training grades by the Paragraph Comprehension ability score.

Further, the discriminant validity also varies with cognitive pattern. The tests with complex decisions correlate more highly with other aptitude measures than the tests with complex decisions, with one exception: Numerical Operations is more highly correlated with Paragraph Comprehension ability when the text representation component is complex.

However, since incremental validity did not vary with cognitive pattern, when added to the ASVAB the lower discriminant validity for tests with the decision as the source of complexity will not lessen predictive validity.

These results indicate that text representation as a source of cognitive complexity in these items lowers the validity of the ability test scores. The comparison of the CAT-PC item bank with the Gates–MacGinitie Test of paragraph comprehension suggests some possible explanations for this effect. The Gates–MacGinitie Test is a standard for measuring paragraph comprehension, particularly as related to reading.

In general, the cognitive characteristics of the CAT-ASVAB differed significantly from the Gates–MacGinitie. For the text representation process, the comparisons indicated that the CAT-ASVAB text contains a lower vocabulary level and larger amounts of syntactic, rather than semantic information. For the decision component, however, the CAT-ASVAB items were more difficult on the reasoning required to falsify or confirm alternatives. As for the text, the Gates–MacGinitie was more difficult on the vocabulary level. These results suggest that the Gates–MacGinitie is a better measure of the semantic aspects of paragraph comprehension while the CAT-ASVAB primarily measures verbal reasoning.

These results suggest that the computerized adaptive version of the Paragraph Comprehension test should minimize the current sources of text complexity as a source of item complexity. This could be accomplished most easily by specifying standards for syntactic complexity, including variables such as the percent of content words and density of connective and modifier propositions. Another approach would be to balance items for text and decision complexity. However, the generally poor showing of tests that are complex on the text counters this as a solution.

These recommendations should not rule out, however, the possibility of increasing text complexity in ways more similar to the Gates–MacGinitie. More complex semantic information may well lead to a paragraph comprehension test that would increase incremental validity further over the over ASVAB subtests. Semantically complexity is currently not well represented in the current Paragraph Comprehension tests.

REFERENCES

Anderson, R. C. (1972). How to construct achievement tests to assess comprehension. *Review of Educational Research, 42*, 145–170.

Booth-Kewley, S., Foley, P. P., and Swanson, L. (1985). *Predictive validation of the Armed Services Vocational Aptitude Battery (ASVAB) Forms 8, 9, and 10 against performance in 100 Navy schools* (Report TR 85–15). San Diego, CA: Navy Personnel Research and Development Center.

Bovair, S., and Kieras, D. (1981). *A guide to propositional analysis for technical prose* (Report No. 8 for Office of Naval Research Contract N00014-78-0509). Tucson, AZ: University of Arizona.

Embretson, S. E. (1983). Construct validity: Construct representation versus nomothetic span. *Psychological Bulletin, 93*, 179–197.

_____ (1984). A general latent trait model for response processes. *Psychometrika, 49*, 175–186.

_____ (1985a). *Component latent trait models of paragraph comprehension* (Final Report for Contract 0855), Research Triangle Park, NC: Battelle Research Institute.

_____ (1985b). Multicomponent latent trait models for test design. In S. Embretson (Ed.), *Test design: Ne directions in psychology and psychometrics.* New York: Academic Press.

_____ (1985). The problem of test design. In S. Embretson (Ed.), *Test design: New directions in psychology and psychometrics.* New York: Academic Press.

Embretson, S. E., Schneider, L., and Roth, D. (in press). The impact of multiple processing strategies on verbal reasoning ability. *Journal of Educational Measurement.*

Embretson, S. E., and Wetzel, D. (1985). *Component latent trait models for paragraph comprehension tests.* Unpublished manuscript, University of Kansas, Lawrence, KS.

Fischer, G. (1973). Linear logistic test model as an instrument in educational reearch. *Acta Psychologica, 37*, 359–374.

Flesch, R. (1948). A new readability yardstrick. *Journal of Applied Psychology, 32*, 221–233.

Kintsch, W., and van Dijk, T. A. (1978). Toward a model of text comprehension. *Psychological Review, 85*, 363–394.

Kucera, H. and Francis, W. W. (1967). *Computational analysis of present-day American English.* Providence, RI: Brown University Press.

Moreno, K. E., Wetzel, C. D., McBride, J. R., and Weiss, D. J. (1984). Relationship between corresponding Armed Services Vocational Aptitude Battery (ASVAB) and computerized adaptive testing. *Applied Psychological Measurement, 8*, 155–163.

Pellegrino, J. W., and Glaser, R. (1979). Cognitive correlates and components in the analysis of individual differences. *Intelligence, 3*, 187–214.

Sternberg, R. J. (1977). *Intelligence, information-processing and analogical reasoning: The componential analysis of human abilities.* Hillsdale, NJ: Lawrence Erlbaum Press.

_____ (1985). *Beyond IQ: A triarchic theory of human intelligence.* Cambridge, MA: Cambridge University Press.

Turner, A., and Green, E. (1978). Construction anduse of a propositional text base. *JSAS Catalog of Selected Documents in Psychology, 3, 58*, MS–1713.

Whitely, S. E., and Barnes, G. M. (1979). The implications of processing event sequences for theories of analogical reasoning. *Memory & Cognition, 1*, 323–331.

Wolfe, J. (1985). *Speeded tests: Can computers improve measurement?* Paper presented at the 1985 Military Testing Conference; San Diego, CA.

4   Computerized Adaptive Testing of a Vocational Aptitude Battery

Martin F. Wiskoff
Mary K. Schratz

INTRODUCTION

For the past 70 years military personnel research has addressed many issues of mental testing and has been recognized as being in the forefront of psychometric measurement and test development. During World War I there was an urgent need to assess the abilities of the large number of recruits brought in under rapid mobilization. The multiple choice Army Alpha—along with its nonverbal equivalent, the Army Beta—were constructed to provide quantitative measures for assigning personnel into schools and jobs, and the era of group intelligence tests was born. World War II brought about a huge personnel selection and classification task which led to the development of the first multiple-ability aptitude batteries.

Today military researchers are spearheading the application of a new mental measurement tehcnology, computerized adaptive testing (CAT). CAT differs from conventional paper and pencil testing in two major respects: tests are administered by computer and they are adapted or tailored to the ability level of the examinee. The capability to tailor the tests is made possible by CAT's underlying theoretical framework called item response theory.

CAT is realtively young, only dating back to the 1950s, yet it has been the subject of extensive investigation, much of it under the sponsorship of military research organizations such as the Office of Naval Research (ONR), the Navy Personnel Research and Development Center (NPRDC), the Army Research Institute (ARI), and the Air Force Human Resources Laboratory (AFHRL).

In 1979, the Department of Defense officially recognized the potential of CAT for improving the process of accessioning personnel in to the armed services, and initiated a program to develop a CAT version of the Armed Services Vocational Aptitude Battery (ASVAB) (Wiskoff, 1981). The ASVAB is administered to all applicants for enlistment and is a major factor in determining military service eligibility and occupational entry.

The CAT-ASVAB system development has evolved into an intricate blending of psychometrics and computer systems concerns, overlayed by studies to insure adequate program documentation for internal and external review. The first section of this chapter will describe the evolution of computerized adaptive testing, including its underlying psychometric theory of test development and the adaptive testing process. This will be followed by a description of the ASVAB, its shortcomings, and the goals of the CAT-ASVAB program.

The program itself is organized into the four domains portrayed in Figure 4.1. This chapter will address in turn the issues of psychometric development, computer system development, implementation studies, and future testing systems. Emphasis will be placed on the considerable advances in knowledge already achieved by the program and the exciting prospects for future psychometric measurement.

Figure 4.1
Paradigm for Developing a Computerized Adaptive Testing Version of the Armed Services Vocational Apititude Battery (CAT-ASVAB)

CAT-ASVAB Program

PSYCHOMETRIC DEVELOPMENT

Item Banks

Test Administration and Scoring

Reliability and Validity

Technical Manual

Scaling and Equating

COMPUTER SYSTEM DEVELOPMENT

Functional Requirements

System Design and Evaluation

Data Flow and Management

IMPLEMENTATION STUDIES

Examinee Attitudes

Public Information and Education

Program Evaluation and Maintenance

Concept of Operations

FUTURE TESTING SYSTEMS

CAT-ASVAB Enhancements

Measuring New Domains

Cross-National Programs

EVOLUTION OF COMPUTERIZED ADAPTIVE TESTING

Item Response Theory

Since the beginning of this century, measurement of human cognitive abilities has often involved the administration of paper and pencil tests to groups of individuals. Such tests have typically been developed through traditional or "classical" test theory methods (Gullicksen, 1950). These methods usually involve the following steps: (1) development of a well-defined set of experimental test questions; (2) administration of the questions to a large group of individuals to obtain necessary information on the usefulness of specific test questions for the intended purposes of the measuring instrument; (3) revision and selection of test questions to formulate a test form which meets predesigned test specifications with respect to content, difficulty level, reliability, and validity; and (4) administration of this form intact to a representative sample of examinees for the purpose of standardization, that is, establishing scales, norms, or "equivalent" scores.

Subsequent administration of the test for operational purposes must employ the test items comprising the complete intact form *in the exact way* that they appear in the standardization test booklet and/or answer sheet. The test must be administered under the same time limits and test administration conditions for which the instrument was normed. All examinees taking a given test form are presented with identical test items and their scores are reported in terms of relative standing among peers.

Measurement method are changing, due in part to the use of high speed, sophisticated computers in research on new psychometric techniques. One such recently developed measurement technique is *item response theory* (Lord, 1980). The promise that item response theory has for applied measurement problems stems primarily from the invariance property for item parameters which holds that: (1) item parameters that are independent of examinee population characteristics may be estimated; and (2) once a set of items has been calibrated, a given individual would receive the same measurement (estimated ability, except for errors of measurement), regardless of the particular subset of items taken. Over the years, measurement experts have noted the desirability of these features for a measuring scale (Gardner, 1966).

In item response theory, the probability that an examinee will respond correctly to an item is expressed as a mathematical function of the parameters for the examinee and for that item only. There have been several alternative item response theory models proposed. However, the three-parameter logistic model has achieved the greatest popularity and use in recent years. The logistic item response curve, giving the probability that a person with ability θ responds correctly to item j, is represented by the following equation (Green et al., 1984):

$$P_j \theta = c_j + \frac{1 - c_j}{1 + \exp(-1.7a_j(\theta_i - b_j))}$$

Green et al. (1984) point out that a can be interpreted as item discriminiability, b as item difficulty, and c as pseudochance level. When θ is very low, the probability of a correct response is c_j; as θ increases, the probability rises from c_j to 1. For high values of a, the rise in probability is sharp, and, for low values of a, the rise is gentle. It is this model, rather than those including one or two parameters only, that has been generally recommended for use with multiple-choice items, where there is an opportunity for the examinee to guess.

Computerized Adaptive Testing

Item response theory provides a theoretical framework for "tailored" or "adaptive" testing. Adaptive testing, using item response theory methods, is a process in which test items are selected for administration to the examinee based on estimates of an examinee's ability calculated from his or her responses to previously administered items. The adaptive testing process is an iterative one involving: (1) determination of an initial estimate of examinee ability; (2) administration of a test item; (3) scoring of the response to the item; (4) determination of a revised examinee ability estimate; and (5) selection of a new item for administration. With the integration of item response theory into the adaptive testing process, examinee ability estimates can be made from both a response to a single item, and from the responses to a series of items. The process allows for the most appropriate set of items to be administered to examinees, and for the scores earned from the administration of different sets of items to be reported on the same scale. The combination of item response theory, adaptive testing, and interactive computer administration of tests is known as computerized adaptive testing (Weiss & Kingsbury, 1984).

There are many critical considerations in the development of a computerized adaptive testing program, particularly when the test involved is an established and high-volume test battery like the ASVAB. As lead laboratory, NPRDC sought the assistance of a panel of psychometric experts in the formulation of an evaluation plan for a computerized adaptive vocational aptitude battery (Green et al., 1982). The plan provides comprehensive guidelines for the psychometric development of CAT-ASVAB, and also addresses relevant computer hardware and test administration issues.

Many of the issues included in the evaluation plan are integral to the development of any battery of ability tests, for example: (1) test content specification; (2) test dimensionality; (3) reliability and measurement error; (4) validity and differential prediction; and (5) equating of ability scales. The development and implementation of a computerized adaptive aptitude battery, however, dictates the need for additional documentation concerning: (1) the formation and use of test item "pools"; (2) adaptive test administration and scoring procedures; and (3) human factors decisions regarding the computerized testing process. Other guidelines for computer-based tests and interpretations have recently emerged (American Psychological Association, 1986), and the *Standards for Educational and Psychological Testing* (American Psychological Association, 1985) are also applicable to this program. Psychometric evidence developed in support of the feasibility of CAT-ASVAB is described later in this chapter; greater detail can be found in the CAT-ASVAB Technical Manual.

CAT-ASVAB PROGRAM

The ASVAB is a paper and pencil administered instrument consisting of eight power and two speeded tests. Figure 4.2 describes each of these ten tests.

The ASVAB is administered as a part of two programs for recruiting and accessioning personnel. The first is directed toward high school students who have exhibited some interest in military occupational entry. The battery is typically given to groups of students, with results made available to test takers, counselors, and military recruiters.

The second program is designed to measure applicants to the military. Testing occurs at more than 950 sites across the United States. Of these, there are 69 permanent locations termed Military Entrance Processing Stations

Figure 4.2
ASVAB Subtests

Subtest	Description	Number of Items
General Science	Knowledge of physical and biological sciences	25
Arithmetic Reasoning	Ability to solve arithmetic problems	30
Word Knowledge	Understanding of the meaning of words	35
Paragraph Comprehension	Ability to obtain information from written passages	15
Numerical Operations	Speed of mathematical computation	50
Coding Speed	Speed of finding a number in a table	84
Auto and Shop Information	Knowledge of automobiles, shop practices, and the use of tools	25
Mathematics Knowledge	Knowledge of high school level mathematics	25
Mechanical Comprehension	Understanding of mechanical or physical principles	25
Electronics Information	Knowledge of electricity, radio principles, and electronics	20

(MEPS), with the remainder being widely dispersed Mobile Examining Team (METS) Sites. Approximately 70 to 80 percent of examinees are tested at the METS; those who decide to enlist will have their final processing (for example), at the MEPS prior to service entry.

A number of scores are obtained from the ASVAB, the most publicized being a "mental level" measure. This percentile index of military trainability is a composite of four ASVAB subtests and is used to determine basic eligibility for enlistment. Numerous other composites of ASVAB are used to evaluate potential for various military technical schools and for occupational entry.

Although the information from ASVAB is critical to the effective training and vocational placement of military personnel, the test and administration process have many shortcomings:

1. ASVAB testing can take as much as four hours with some applicants. In many instances this creates scheduling problems in high school and recruiting environments.

2. Any single unspeeded test in the battery has relatively few items and these have been calibrated to provide maximum measurement power in the center of the ability distribution. As a result, the precision obtained for high and low ability applications is often inadequate for certain recruiting and placement decisions.

3. Paper and pencil tests are susceptible to breaches of security such as theft, compromise, and coaching along with human errors in scoring and recording. The ASVAB suffers the same problems.

4. The widely dispersed and variable nature of ASVAB administration introduces the potential for situational testing inconsistencies. This can lead to significant differences in applicant scores, particularly in the speeded subtests, depending on where and how a person is tested.

5. Development of replacement ASVAB forms requires a long lead time, it is expensive and very time-intrusive on applicants and service recruiting personnel.

6. ASVAB shares a problem with other paper and pencil test batteries, in that the use of test booklets severely limits the range of human aptitudes and abilities that can be assessed. For example, it is virtually impossible to measure information processing or stimulus tracking, both critical job skills in the military.

These problems present serious difficulties to military managers concerned with the most effective utilization of enlisted personnel. The most immediate CAT-ASVAB program goals in response to these concerns are to develop a system for use in military recruiting that incorporates the following attributes (McBride, 1982):

1. On-line interactive administration of personnel tests using automated display and response media.

2. Dynamic tailoring of test difficulty to each examinee's ability contingent on performance at earlier stages of the test. This tailoring will reduce ASVAB test duration by 50 percent or more and will improve measurement precision at the extremes of ability levels.

3. Administration of a unique tailored sequence of test items to each examinee. Each test is drawn from a very large bank of test items. The size of the item bank will effect a significant defense against test compromise and coaching.

4. Replacement of all printed test material by electronic media, thus eliminating printing and storage costs of test booklets and answer sheets. This replacement will also enhance test security because no printed materials will be available for theft.

5. Computerized scoring, score conversion, score composite computation, and score recording, thus eliminating erroneous personnel record data attributable to clerical errors in manual scoring and recording.

6. On-line administration of experimental replacement test questions to examinees, which will reduce the lead time needed to construct replacement test forms.

Future program goals are to: (1) evaluate the potential of computer-administered tests for enhancing ASVAB prediction of an applicant's ability to perform in military training and operational performance situations, and (2) evaluate the feasibility of administering CAT-ASVAB in high schools and junior colleges.

The program is being performed in two major stages. The first, called Accelerated CAT-ASVAB Project (ACAP), is evaluating the feasibility of computerized adaptive testing technology in a limited number of MEPS and METS locations. ACAP is using off-the-shelf, commercially available microcomputer equipment and designing software to meet system functional requirements. An early phase, called the experimental system, was developed on Apple III computers. The current ACAP program, which will be described later, employs a Hewlett-Packard Integral computer. The second stage will utilize the lessons learned from ACAP to install a nationwide system for assessing applicants for military service.

Under the sponsorship of the Office of the Secretary of Defense, the Department of the Navy has responsibility for the CAT-ASVAB program. NPRDC, San Diego, is designing all program components with assistance from the other military personnel research laboratories and with supporting contracts. Technical guidance is provided by the Defense Advisory Committee on Military Personnel Testing, whose members are recognized experts in the field of personnel measurement.

CAT-ASVAB PSYCHOMETRIC DEVELOPMENTS

Item Banks

The development of a paper and pencil test usually involves creating a set of test items covering the ability range of the population for which the test is designed with all test takers being administered all items. Accordingly, given realistic testing time constraints, the greatest number of items are written to be of appropriate difficulty for the greatest number of people, that is in the middle of the ability range. In contrast, the adaptive testing process requires that examinees be administered only those items suited to their ability level; consequently, large pools of test items must be developed containing more items than are needed for a paper and pencil test and spanning a wide range of difficulty. If the adaptive test battery is to replace a traditional instrument as is the case for CAT-ASVAB, domain specifications must often be revised to accomodate these increased numbers of test items and yet cover the same skills.

The item response theory model employed in CAT-ASVAB requires that certain conditions be met and reflected in characteristics of the item pools. One of the conditions for application of the three-parameter logistic model is that each aptitude test area must be unidimensional. To fulfill this requirement, each test item should measure the same unitary construct, in addition to its specific and error components. Green et al. (1984) point out, however, that, while item response theory is based on unidimensional items, empirical results show that the model is suitable even when the assumption is violated, as long as there is one dominant dimension.

The unidimensionality condition of the item response theory model was taken into account in planning the development of CAT-ASVAB item banks. For example, while previous P&P-ASVAB test forms have combined the Automotive and Shop Information test items into a single test, a decision was made to develop enough CAT-ASVAB items of each type to allow for the creation of two separate item pools in these aptitude areas. Consideration has also recently been given to "balancing" test content of physical and life science items in the administration of the CAT-ASVAB General Science test, by fixing the administration of prespecified proportions of items of each type in an examinee's test. The need for this procedure stems from a recommendation made by Green et al. (1982) who point out that proportional representation of the General Science content will, as for P&P-ASVAB, tend to balance out performance differences between subpopulations on identifiable item clusters for the test as a whole.

The condition of unidimensionality also implies local independence, which means that the probability of success on all items equals the product of the separate probabilities of success (Lord, 1980). In other words, the probability of success on an item depends on the three item parameters, on examinee ability θ, and on *nothing* else. Green et al. (1984) point out that item types involving several questions pertaining to the same stimulus violate this condition. The Paragraph Comprehension test of P&P-ASVAB has traditionally consisted of a format involving several test items referring to a single reading passage. The Paragraph Comprehension tests of the experimental system and ACAP have been modified to contain short paragraphs with only one question. This format also minimizes difficulties with respect to the length of the paragraphs a computer screen can display.

Item pools developed for a computerized adaptive test battery may also differ due to the computer presentation mode, as noted above for paragraph comprehension items. The creation of items involving complex graphic displays (for example, in the Mechanical Comprehension aptitude area) must conform to computer display resolution limitations. It is encouraging to note that our work has shown that most graphic display items, except those of extreme complexity, can be reproduced faithfully on the computer screen.

Computer administration also affects the development of speeded tests. The Numerical Operations and Coding Speed tests are constructed so that items do not differ in difficulty; the probability of a correct response is close to 1.0, assuming that the examinee reads and responds to an item before a test limit is reached. While adaptive testing is not feasible for such tests, they can be computer-administered. Choices must be made with respect to the number of items to be displayed per computer screen, the total number of items and time limit, if any, for the test, and the final scoring method. In addition, the norming of such tests will depend on the computer hardware and software that is used.

The major stages of adaptive item bank development may be described as follows:

1. *The content of each aptitude area must be clearly specified.* For an initial CAT-ASVAB item bank, specifications were developed by surveying textbooks from junior and senior high schools, and from vocational technical schools. More than 3,600 test items were written to cover measurement objectives represented by the content specifications; about 400 test items were written for each of nine adaptive power test aptitude areas (Prestwood et al., 1985).

2. *Item reviews by expert judges should be conducted to evaluate the content, format, clarity, and appropriateness of test items.* Such reviews can provide item-by-item suggestions for improving the accuracy of the question as stated and for ensuring that the test material is appropriate for the population. Schratz (1986) describes a formal review of the content of potential CAT-ASVAB items and a P&P-ASVAB reference form by expert judges (test developers and educators). Reviewers pointed to similarities between the CAT-ASVAB item bank and the items of the P&P-ASVAB reference form, and provided recommendations for future CAT-ASVAB items. A second group of expert judges (minority group educators and test developers) reviewed potential CAT-ASVAB items to ensure that the material was as free as possible of that which could be considered objectionable or unfair to minority groups and to men and women; recommendations for improving the test items were also provided.

3. *When sufficient numbers of items covering measurement objectives have been written, reviewed, and modified, it is advisable to pretest them to obtain preliminary information on their difficulty and validity.* The 3,600 potential CAT-ASVAB items were "pretested," in paper and pencil booklets, in a sample of military recruits. Item level statistics were generated and, based mainly on item difficulty indexes with consideration given to test content, approximately 200 items per aptitude area were selected for further development (Prestwood et al., 1985). In addition, it was necessary to supplement each aptitude area with new items designed to be easier than any of those items originally intended for calibration.

4. *Steps must next be taken to determine the item parameter estimates necessary for the computerized adaptive testing process.* Green et al. (1984) have recommended that, for item parameter estimates obtained via the three-parameter logistic model, a sample of 2,000 cases for each item be considered adequate. The CAT-ASVAB items which remained after screening the pretest data, approximately 2,100 in number, were administered in paper and pencil test booklets to military applicants. Item responses for approximately 2,500 examinees per test item were collected for calibration purposes, and estimates of item parameters were determined using the computer program ASCAL (Prestwood et al., 1985). The item parameter estimates determined in this way are, however, considered preliminary until further research is conducted to verify their accuracy under computer administration.

Collection of the amount of item level data required to support item calibration via the three-parameter model provides a reservoir for future research. Schratz (1986) has described an item factor analysis conducted of this data, using the sophisticated item response theory technique known as marginal maximum-likelihood estimation, as implemented in the computer program TESTFACT (Bock, Gibbons, & Muracki, 1985). This data can also be utilized in exploring research questions pertaining to differential item functioning between groups, that is: Do the item trace lines differ for subpopulations? In our work, we have found that preparing an item bank for use in a computerized adaptive testing program also involves looking for new techniques and procedures: (1) to determine item parameter estimates for future items intended to refresh, or replace, those in existing pools; (2) to detect those instances where item parameters "drift"; and (3) to create improved measures through the use of new item types.

Testing Administration and Scoring

The introduction of a computerized adaptive test battery to an existing paper and pencil testing program brings many changes with respect to test administration and scoring procedures. While P&P-ASVAB testing requires that a test proctor keep careful track of the time for administering each test and be well versed in communicating test instructions to a group of examinees, these functions become part of the role of the CAT-ASVAB microcomputer delivery system. While P&P-ASVAB also involves hand and/or machine scoring of an answer sheet once the test session has been completed, the CAT-ASVAB delivery system will dynamically select and score each item and, via item response theory methods, update the ability estimate of the examinee in the aptitude area tested. Rafacz (1986) provides a description of the responsibilities of the test proctor in support of ACAP, and Jones-James (1986) describes the design and development of software automating the test administration and scoring process. CAT-ASVAB test administration and scoring procedures differ for: (1) power tests; (2) speeded tests; and (3) test items included for experimental purposes. Test administration also currently involves providing sample items in each aptitude area for practice purposes. The psychometric procedures for CAT-ASVAB test administration and scoring are described below.

The power tests are administered adaptively and scored through item response theory methods. Current CAT-ASVAB procedures perform the following functions: (1) assigning an initial ability estimate to the examinee for the aptitude area tested, that is, the mean of the latent ability distribution $(\theta = 0)$; (2) selecting an item for administration from those items available at that estimated ability level, that is, an "infotable" is used to select the most informative item, in conjunction with a procedure for controlling the rate of item usage (Sympson, 1985); (3) scoring the response to the last item and updating the examinee ability estimate based on the responses to previous questions in that aptitude area, that is, by using Owen's Bayesian estimation procedure (1969); and (4) terminating each test after a fixed number of items have been administered and each subsequent ability estimate determined, that is by repeating steps (2) and (3) above. Consideration is being given to alternative procedures for determining a final ability estimate for the examinee in the area tested; a Bayesian model estimator of theta is the current favorite.

The speeded tests, while computer-administered, are not adaptive and item response theory methods are not appropriate for scoring. A termination criterion must still be determined for these tests; for P&P-ASVAB, this criterion is a fixed time limit and an examinee's score is determined by the number of items answered correctly. Recent research (Wolfe, 1985; Greaud & Green, 1986) has examined methods for administering and scoring speeded tests. This work has led to the use of the following procedures in the CAT-ASVAB delivery system for ACAP: (1) administering one Numerical Operations item per computer screen, similar to the procedure followed for CAT-ASVAB power tests, and administering several Coding Speed test items per screen, similar to the format of the P&P-ASVAB test; (2) administering the same number of CAT-ASVAB test items in the same time limit as for P&P-ASVAB, although responding in a computerized mode is much quicker; and (3) using a rate score, that is, the geometric mean response time across computer screens.

Items included in CAT-ASVAB for experimental purposes are, of course, computer-administered and scored as right or wrong; they are not, however, administered adaptively, or used in determining an ability estimate by item response theory or other methods. Such items are included primarily to further research on: (1) estimating new item parameters, and verifying existing ones, obtained under the same or different modes of administration; and (2) pretesting new types of test items for potential inclusion in CAT-ASVAB item banks.

Operating characteristics of future CAT-ASVAB systems will contain the findings from current research pertaining to measurement precision, rates of item usage, procedures for the calibration of new test items, and new item types.

Reliability and Validity

Fundamental to the evaluation of any measuring instrument is the examination of estimated measurement error associated with resulting test scores and the validity of test scores for their intended purpose. The identification of major sources of measurement error and the estimation of the sizes of the errors resulting from these sources is generally known as test reliability. In classical test theory, reliability is technically defined as the ratio of true score variance to the observed score variance in the population of persons from which the examinees are assumed to be randomly sampled

(Green et al., 1984). The correlation, ρ, can be estimated directly by alternate-form (or test-retest) reliability and, indirectly, by internal-consistency (or split-half) reliability. Reliability is estimated as a number between 0 and 1; a more meaningful expression of reliability for score interpretation is, however, the standard error of measurement, in the metric of the reported score.

While in classical test theory a single reliability coefficient, and a single standard error of measurement, is typically reported pertaining to all score levels, measurement error is expressed as a function of ability in item response theory-based testing. For an adaptive test, measurement error also depends on the stopping rule used. The test can be ended at a specific target error variance for all ability levels, or some other rule may be used such as administering a fixed number of items. In the latter case, measurement error will vary with ability level and with the number of specific items administered.

Both the experimental system and ACAP employ a stopping rule based on a fixed number of items. In such cases, measurement error, in the metric of the reported score, must be determined for various levels of ability. Since the standard error varies not only as a function of ability but also as a function of the particular adaptive test to which the examinee responds, the range of standard errors for these ability levels must also be determined. Measurement error must be computed for all subtests of the CAT-ASVAB battery, and for the composite scores (combinations of subtest scores) reported for selection and classification purposes. Further information on the reliability of the CAT-ASVAB battery is obtained by examining error variance due to repeated administration of different test items by the adaptive testing algorithm; alternate-form reliability studies and test-retest reliability studies can be conducted using different item pools. Such studies are the only indication of reliability appropriate for the speeded tests. The experimental Apple III system has yielded some information on test-retest reliabilty (Moreno, 1985). In a study conducted among military recruits, CAT-ASVAB test-retest reliabilities, based on alternate test forms for which a fixed number of items were administered, compared favorably to those of P&P-ASVAB. Information on the reliability and measurement error associated with versions of CAT-ASVAB intended for operational use must come from further empirical studies.

The psychometric property cited as most important in evaluating any measuring instrument is test validity (American Psychological Association, 1985). According to the *Standards for Educational and Psychological Testing*, validity refers to the "appropriateness, meaningfulness, and usefulness" of specific inferences made from test scores. The *Standards* point out that there are a variety of inferences that may legitimately be made from scores produced by a given test, and that there are also many ways of accumulating evidence to support any particular inference.

Since the P&P-ASVAB is used in the selection and classification of military applicants, its "predictive validity," or the relationship between performance on the test and on the criterion, is of primary importance. The primary criteria for evaluating the predictive validity of P&P-ASVAB have been training school scores (United States Military Entrance Processing Command, 1984). The most recent test mannual for P&P-ASVAB provides some reasons for the appropriateness of these criteria: (1) there are, at present, no uniform measures of job performance across the services; (2) those individuals who do not successfully complete training cannot do the job. The services establish the content of training courses based on objective occupational analyses for each of their specialties. Since there are a number of occupations found in each service, the accumulation of validation samples of sufficient size for statistical stability is a time-consuming project. Nevertheless, predictive validities for more than 100 occupational specialties are reported for the latest forms of P&P-ASVAB in the test manual.

These studies are relevant in considering the predictive validity of CAT-ASVAB as a replacement for the P&P-ASVAB. Since the collection of data for predictive validity studies is a process which continues throughout the life of the test and is not available until after a form of the test has been in operational use for some time, no such data can be reported for ACAP. However, studies described by Day, Kieckhaefer, and Segall (1986) have indicated that the experimental Apple III system shows similar predictive validity to the P&O-ASVAB forms. These results were based on 7,513 recruits to the armed services who were scheduled for later attendance at one of several technical schools. Training curricula and individual performance data were gathered from the training schools of 23 occupational specialties across the four services. The results of these preliminary studies indicate that the experimental CAT-ASVAB subtests and selector composites were as valid as those for P&P-ASVAB; where differences were found, they were most often in favor of the experimental CAT-ASVAB. Studies to validate the operational CAT-ASVAB battery are being planned.

Examination of the content of item pool for each subtest can provide further evidence for the validity of the CAT-ASVAB battery. The key issue here is, of course, how well the item pool represents the domain of cognitive abilities or skills comprising the aptitude area. A brief description of the reviews by expert judges, providing a verification of item level content specifications and a comparison, at the test level, of the content of the initial CAT-ASVAB item bank and that of a P&P-ASVAB reference form, is provided in the section on Item Banks.

Evidence that the items developed for CAT-ASVAB measure the same trait or construct as P&P-ASVAB provide further support for the validity of inferences made from CAT-ASVAB test scores. Some preliminary investigations of the underlying factor structure of the P&P-ASVAB and the experimental Apple III system have been conducted (Day et al., 1986; Martin, Park, & Borman, 1986). These studies have basically replicated the four-factor structure of the P&P-ASVAB (verbal, quantitative, technical, and speed).

Finally, modern methods of structural equation modeling can be useful in evaluating relationships between the two test batteries; models of this sort have been described by Joreskog and Sorbom (1986). Such models can also provide evidence of the effects on scores due to mode of test administration, if such effects exist. Some early work, using data collected with the experimental Apple III system, has been done in this area (Martin et al., 1986); further study should involve batteries intended for operational use.

Technical Manual

Integral to the introduction of a large-scale testing program is the compilation and documentation of technical data to supports its operational use for intended testing purposes. To this end, a comprehensive CAT-ASVAB Technical Manual covering psychometric and system development progress is in preparation (Schratz, 1985). Chapters of this Manual contain the following material:

1. *An introduction to the CAT-ASVAB program.* The origins of mental testing in the military are presented, including the development of the Army Alpha and precursors to the modern Armed Services Vocational Aptitude Battery. The dilemma of group versus individualized testing and the concept of adaptive testing is introduced, along with a statement of the military's early interest in item response theory and in adaptive testing.

2. *Background.* The function and administrative aspects of P&P-ASVAB in military testing and the motivation for CAT-ASVAB is addressed. This background chapter also provides a discussion of measurement issues in the transition from the paper and pencil test battery to the computerized adaptive battery.

3. *Functional characteristics of the delivery system.* The concept of a local CAT-ASVAB network is introduced and data flow and communications in the operational environment are described. Physical and performance characteristics of the ACAP system are highlighted.

4. *Psychometric bases of computerized adaptive testing.* The basic concepts of item response theory and a discussion of the assumptions of the theory as they pertain to the adaptive testing process are included. Procedures for linking item parameter estimates from the same mode of administration and those studies concerned with linking estimates from different modes of administration are addressed.

5. *Development of item pools to support the adaptive testing process.* Requirements imposed by adaptive testing are delineated, as well as those requirements imposed by comparability with the P&P-ASVAB battery.

6. *Testing algorithms.* Included in this chapter are the basic concepts of adaptive testing, procedures for the administration of power tests, studies of system operating characteristics for adaptive tests, and procedures for the administration and scoring of speeded tests.

7. *Scaling and equating of the computerized adaptive test battery to the paper and pencil version of the battery.* Equating methods and an equating data collection design for ACAP are described.

8. *Reliability and measurement precision.* The concept of measurement error is discussed for subtests and composites. Studies comparing the reliabilities of CAT-ASVAB with those of the paper and pencil battery are delineated.

9. *Validity of the CAT-ASVAB battery.* Specific studies addressing content, construct, and criterion-related validity are highlighted and threats to validity, resulting from multidimensionality and sensitivity concerns, are outlined.

The Technical Manual for CAT-ASVAB is envisioned as a living document that will continue to grow as the program advances, more technical data are accumulated, and psychometric/systems advances are made.

Scaling and Equating

The introduction of the CAT-ASVAB test battery to the military accessioning environment will be gradual. During the transition period, military applicants could be tested by either the P&P-ASVAB or the CAT-ASVAB battery. Therefore, it is necessary to establish "equivalent" or comparable scores on both batteries. A computerized adaptive and a conventional test can never be equated in a strict sense, in that the tests are not of equal precision. However, Green et al. (1984) contend that equivalence of expected scores on the tests is sufficient.

The determination of "equivalent scores" will involve the collection of empirical test data for both test batteries among examinees of the intended test-taking population, in the military entrance processing environment. The equating data collection design developed for ACAP involves testing three random samples of approximately 2,500 military applicants with one of the three test forms to be equated (CAT-ASVAB Form 1, CAT-ASVAB Form 2, or P&P-ASVAB Form 8a). The equipercentile methods (Angoff, 1971) employed to place new forms of the P&P-ASVAB on the same scale as a reference form of the battery will be used to scale CAT-ASVAB and P&P-ASVAB. A major objective of this effort is to place the "equivalent" scores on the same scale as scores determined in a reference population of American youth, tested with P&P-ASVAB Form 8a in 1980 (Department of Defense, 1982). The equating will be cross-validated during the initial operational test and evaluation when the scores earned on the CAT-ASVAB battery will be the operational scores of record.

Equipercentile equating of CAT-ASVAB to P&P-ASVAB will yield "equivalent scores" for eight power tests (GS, AR, WK, PC, AS, MK, MC, and EI), for two initial forms of ACAP separately. CAT-ASVAB ability estimates will be equated to raw and standard scores on P&P-ASVAB Form 8a. (The equating of the Automotive and Shop Information tests will actually require steps beyond those necessary for the other aptitude tests.)

Equating the two speeded tests, Numerical Operations and Coding Speed, of the CAT-ASVAB battery to these same tests on the P&P-ASVAB battery will also involve equipercentile methods. However, substituting mean response time for the traditional number-right scores for ACAP will require special attention.

Composites such as the Armed Forces Qualification Test (AFQT) and those used by the services for selection and classification purposes, will also be equated via equipercentile procedures. For the present, CAT-ASVAB composites will be formed of the same subtests comprising the like-named composite of the P&P-ASVAB, and used for the same purpose.

CAT-ASVAB COMPUTER SYSTEM DEVELOPMENT ISSUES

Functional Requirements

A computer system to support the nationwide administration of CAT-ASVAB must meet stringent criteria to insure the availability of testing when required and the accuracy of measures which are obtained. A set of nine major systems criteria have been formulated:

1. *Performance.* The system must automate all current paper and pencil ASVAB functions and future computer testing requirements. System response time to any examinee input cannot exceed two seconds, independent of the system load, for example, number of simultaneous responses to multiple examinations. The display must clearly present both text and pictorial matter with a resolution of at least 512×256 pixels and with the availability of multiple font styles. The display screen must be at least 8-inch horizontal by 4-inch vertical. The design should allow for high-speed electronic data transmission and include appropriate interfaces with the existing MEPS reporting system and future research and maintenance facilities. Computer software design should employ a "top-down" structure, using a high-level language.

2. *Suitability.* The system must be capable of operating in environments where paper and pencil tests are given. Accordingly, there must be no special requirements for temperature or humidity control, for highly trained operators or major site modifications. Additionally, the system must be transportable to allow for test administration in the variety of MET sites.

3. *Reliability.* To support an operational recruiting function, the system must be as available for testing of applicants as under a paper and pencil mode, with sufficiently reliable hardware and backup equipment and/or failure recovery procedures to conduct and complete testing when scheduled. Reliability and availability treshholds of 99.9 percent are the design goals. The system must also be capable of restarting at the point of the first noncompleted test with no loss of data if there is an examinee testing station failure (goal of less than 1 test per 1,000). Failure of any examinee or test administrator station must have no operational impact upon the performance or availability of any other station.

4. *Maintainability.* The hardware and software must incorporate self-diagnostic logic which can be easily understood by test administrators under the assumption that there will be no technically trained technicians on site. Maintainability design must include an integrated life-cycle logistics support system. Software maintenance must be supported by sufficient built-in-test procedures.

5. *Ease of use.* This requirement is critical to the acceptability of the system by test takers, administrators, and the public. Tutorials must be available for

inexperienced computer users and all set-up and administration procedures are to be unambiguous and clearly documented. Consideration must be given to human-computer interfaces, for example, screen display legibility/resolution and examinee response medium.

6. *Security*. System design must incorporate sufficient protection against unauthorized review of test item banks and examinee records. There must be no possibility of making printed copies of test questions and system components should be protected from theft and unauthorized use. The item sequence should be unpredictable to minimize coaching and there should be system access controls and audit trails.

7. *Affordability*. The ten year life-cycle costs and benefits of the CAT-ASVAB sytem must be competitive to that of the paper and pencil ASVAB. This can best be accomplished by developing a concept of operational use that maximizes employment of equipment and utility to users.

8. *Flexibility/Expandability*. Future CAT-ASVAB program goals must be accommodated by incorporation of the capability to administer a wide range of tests under different operational scenarios. For example, there must be a programmable, high-precision clock for measurement of response latency, and provision for using a variety of input peripherals such as a mouse, keypad, keyboard, and devices used in psychomotor testing.

9. *Psychometric acceptability*. The measurement precision of CAT-ASVAB must be equal to or superior to that of ASVAB. Scores from both versions of the battery need to be equated so as to be interchangeable. In general, the system must meet stringent professional test standards and be adequately documented.

System Design and Evaluation

The functional requirements described above have served as guidelines in the consideration of alternative system designs for ACAP. The evaluation of three alternative local CAT-ASVAB network designs has been described by Tiggle and Rafacz (1985). Their recommendations resulted in the choice of a local network with an entire test item bank stored in the random access memory of the examinee test station. This design was considered favorable to those candidate designs storing item banks on removable media or on a central file server; for example, sensitive information will be erased when power is removed and system response time is at a minimum for *any* examinee input.

The selection of a network design drove the decision with respect to the hardware selected for use in ACAP (Rafacz, 1986). For the examinee testing station, the components may be briefly described as: (1) a Hewlett-Packard (HP) Integral PC; (2) an examinee input device developed from the current HP Integral keyboard; (3) a dust cover to protect the station during travel; and (4) a power cord and a network cable stored in the dust cover. For each station, the total computer program directly addressable RAM memory is at least 1.5 megabytes, and the total weight of the package is 25 pounds. The network consists of no more than 30 such stations, monitored by a test administrator station.

The test administrator station consists of the following two packages: (1) a package with the HP Integral PC (with no examinee input device), a printer, a RAM expansion box board, a full keyboard, and a power cord and several cables stored in the dust cover; and (2) a package including the RAM expansion box unit with 3 megabytes of RAM installed in three of the five available expansion slots, a cable, and a power cord. The first of these three packages is 26 pounds and the second is 21 pounds. Finally, there is a package containing peripheral equipment for use in the testing session, such as those discs containing the testing data, power strips, extension cords, etc. Rafacz (1986) has also described the duties of the test administrator in conducting the session.

All of the software developed for the ACAP computer system is in the "C" programming language. This language was chosen because it is native to the UNIX System V operating system installed in ROM of the HP Integral PC, and because it has certain characteristics which are important in software development, performance, and testing (Rafacz, 1986). Desirable features of this language include: (1) support of structured programming; (2) portability; (3) execution speed; (4) concise definitions and fast access of data structures; and (5) real-time system programming. A top-down structured design approach was used in software development. Jones-James (1986) has described the design and development of the ACAP test administration software using this top-down design structure, whereby modular division of tasks enhances the performance and maintainability of the software.

There are two basic modes of operation for the ACAP delivery system: "networking," the predominant mode, and "standalone," the back-up (failure-recovery) mode. Through modular software design, functions for networking and standalone modes of operation can be isolated as distinct modules. Jones-James (1986) describes a "Bootup" program, part of the test administration software, through which the mode of operation is identified and, once this is determined, a module is called to load the examinee test station software from either the network or directly from ACAP system discs. The examinee test station softward performs the following functions: (1) accepting test data; (2) identifying the examinee; (3) administering the CAT-ASVAB test battery; and (4) transferring the examinee test results to the test administrator's station. The software for the test administrator's station supports the following functions: (1) configuring the local network; (2) randomly assigning test forms to examinee test stations; (3) downloading data through the network; (4) assigning examinees to test stations; (5) monitoring examinees at the test stations; (6) collecting examineee test results; and (7) recording all examinee data onto a single disc for communication to a parent MEPS via registered mail.

The hardware and software described above for the ACAP delivery system was field-tested during October/November 1987, at the MEPS in

San Diego. The CAT-ASVAB battery was administered to military applicants, followed by a written questionnaire or a verbal interview pertaining to the examinee's perceptions of the test. Concurrently, for research purposes only, a special sample of low-ability high school students was also tested using ACAP. The data collected in the field test confirmed the ease of use of the hardware and software currently comprising the ACAP delivery system. As dictated by the field test, modifications of the hardware and software will be made in advance of collecting data for the purpose of "equating" CAT-ASVAB test scores and those scores derived from P&P-ASVAB, as described in the previous section of this chapter.

Data Flow and Management

A "data handling computer" will play a major role in the collection and distribution of data at the military entrance processing stations, the United States Military Entrance Processing Command (USMEPCOM) headquarters, and the CAT-ASVAB Maintenance and Psychometric (CAMP) Facility (as described in next section). This process is envisioned as involving the following steps: (1) at the completion of a CAT-ASVAB test session, a Datadisc will be created at the test administrator's station containing session data for all examinees tested and forwarded to a MEPS; (2) a Data Prep module reads the disc, separates USMEPCOM operational test data from the CAMP Facility research data, and stores all data on the data handling computer; (3) the operational test data are then transmitted, via a communications link, to an existing minicomputer system at the MEPS; (4) a CAMP tape module is used to consolidate all examinee test data collected within the administrative segment of this MEPS onto a single tape cartridge which is then sent to the data handling computer at USMEPCOM headquarters; and (5) the data handling computer at headquarters consolidates all tapes received from the various MEPS onto a single CAMP tape which is then sent, via registered mail, to the CAMP Facility (Folchi, 1986).

CAT-ASVAB IMPLEMENTATION STUDIES

CAT-ASVAB will ultimately affect the lives of millions of this nation's young people, their parents, and the public at large. Studies are being conducted to enable test takers and major segments of the population to have an opportunity to express opinions and introduce constructive changes as the program evolves toward implementation. Additional studies are addressing total life-cycle management, which includes system evaluation and maintenance and the assessment of economic and other considerations which would influence the manner in which the system is deployed.

Examinee Attitudes

While computers have permeated almost every aspect of our daily lives, there nevertheless are significant differences among individuals in their computer knowledge and exposure. Recognizing these differences, the psychometric community has expressed interest in determining attitudes toward taking tests via computers especially among those with little previous familiarity.

Early speculation was that computer usage generates anxiety which can negatively impact on user attitudes and performance (Nilles et al., 1980). This however was not substantiated by Schmidt, Urry, and Gugel (1978), who found that examinees completing a computerized adaptive test similar to the Civil Service entrance examination exhibited overwhelmingly positive attitudes toward the testing situation. More recently, Ward, Kline, and Flaugher (1985) found very favorable reactions by large numbers of students, faculty, and staff to an adaptive version of the College Board Computerized Placement Test.

Several attitudinal studies have been conducted specific to CAT-ASVAB. A preliminary look at user acceptability, with a sample of 227 male Navy recruits taking the CAT-ASVAB experimental battery, indicated that 95 percent enjoyed taking the test by computer (Mitchell et al., 1983). A related survey with 1,237 male recruits from various military services found similar positive attitudes (Hardwicke and Yoes, 1984).

In the most definitive study to date, Kieckhaefer, Segall, and Moreno (1985) compared attitudes of 619 Navy recruits toward CAT-ASVAB versus paper and pencil ASVAB. Recruits were randomly assigned to take one or another test first, with a nine-item questionnaire administered subsequent to completion of the first test. Respondents taking the experimental computerized version of ASVAB indicated they were less tired, less pressured, experienced less eye fatigue, felt the instructions were clearer, and felt better about taking the test. There were no significant differences between the groups on feelings about test fairness, question difficulty, and their anxiety in taking the test.

A recent study on military examinees in West Germany (Wildgrube, 1985) found that a CAT application was preferred to a paper and pencil version even among those with no experience with home computers or video games.

Public Information and Education

The visible differences between the current paper and pencil and future computerized adaptive systems of test administration will raise questions in the minds of many groups. Preliminary indications are that congressional concerns will focus on the costs versus benefits, potential impact on minority/ethnic groups, and the technical soundness of the system to avoid misnorming issues that have occurred with ASVAB. Military service interests revolve around the performance of CAT-ASVAB, for example, ability to accomplish the job of testing applicants, quality of measurement, details of operational use, and compatability with existing military applicant processing and reporting systems. Minority, ethnic, and women's groups have expressed strong interest in learning whether CAT-ASVAB will treat subgroups fairly and whether there would be adverse impact. Applicants and parents will be most concerned with the accuracy and proper use of test results and the fairness of testing via computer.

The needs of these diverse groups can be best accommodated through the generation of tailored multimedia information programs. Materials to be prepared include congressional point papers, minority and user seminars, applicant and information brochures, video presentations, general interest articles, and research papers. The unique aspect of this approach is the concerted effort to go public with information; the philosophy being that a well-informed nation will appreciate the inherent advantages of computerized testing and will not react negatively such as might happen if CAT-ASVAB suddenly appeared without prior knowledge.

Program Evaluation and Maintenance

All large-scale testing programs routinely conduct systematic evaluations of examination results and processes; in the last 20 years it has become a matter of public interest. There was great speculation in the 1960s and 1970s as to the causes of the continuing score declines on the Scholastic Aptitude Test (SAT) (Wharton, 1977). The consternation in the 1980s, accompanying the disclosure that there had been inadvertent misnorming of the ASVAB, testifies to the need for close scrutiny of testing programs and for the capability to rectify inconsistencies (ASVAB Working Group, 1980).

Since its introduction in 1976 as the replacement for separate military service screening and classification tests, the ASVAB has been revised on an approximately four-year cycle. Replacement forms have been developed, reviewed, equated to existing forms, normed, and then evaluated in an operational setting, prior to full acceptance of the scores.

One of the major advantages offered by CAT-ASVAB is the potential for more accurate and timely design of replacement forms, at lower cost, and with less disruption of operational testing. Currently, new test forms must be tried out along with the operational ASVAB under circumstances where applicants are aware they are taking an experimental instrument. This has sometimes caused doubts as to applicant motivation and therefore the accuracy of experimental data. Under CAT-ASVAB, experimental test items can be interspersed among operational ones making them transparent to test takers. This process of "on-line calibration" of test items will be a major feature of future computerized testing systems.

There are many additional opportunities afforded by computerized test administration for program appraisal, new test development, and research. The relatively uncharted nature of this world, however, generated the requirement for establishing a concept and operational procedures for a CAT-ASVAB Maintenance and Psychometric (CAMP) Facility (Vale, Lind, & Maurelli, 1986). The overall mission of the CAMP facility can be summarized as five distinct functions:

1. serve as the interface for revising CAT-ASVAB tests and introducing modifications into the operational system;
2. support USMEPCOM in the distribution of test and system revisions and in the collection of data from field sites;
3. operate as the central repository for all data collected on the CAT-ASVAB sytem;
4. develop automatic quality assurance procedures and perform analyses to detect/correct systems problems;
5. conduct state-of-the-art research leading to significant improvements in future versions of CAT-ASVAB.

While the functions of CAMP will roughly parallel those required to maintain paper and pencil tests, the nature of the operations will be very different. CAMP is envisioned as having the capability to store vast amounts of data required to conduct sophisticated analyses heretofore impossible. For example, it will be feasible to: address questions such as the impact of item exposure on item characteristics; promptly detect item or content area compromise; monitor the internal consistency of an individual examinee's responses; analyze response latencies; and perform demographic analyses.

The CAMP Facility will utilize the services of an on-site mainframe computer accessed via microcomputers to perform the daily system evaluation and mainenance. It will also contain a computer laboratory with operational CAT-ASVAB delivery system equipment, which will be used by NPRDC and other service researchers, for development and evaluation of experimental tests.

Concept of Operations

The consideration of CAT-ASVAB as a nationwide computerized testing system raises the logistical issue of where system components will be located. It also introduces concerns as to the costs and benefit associated with the system relative to paper and pencil testing. Because of the obvious relationships between siting strategies and economics, these matters are being addressed by evaluating alternative concepts of operations in terms of their benefits and economic consequences.

One possible approach is to simply attempt to emulate paper and pencil testing with computerized testing, that is, for all of the 950+ fixed and mobile locations where ASVAB is currently administered a CAT-ASVAB equivalent would be made available. While such a system would minimize the operational impact on the current testing process, it would require a large investment in computer equipment and in site modification for test administration and computer storage, and would not be very efficient in the time-sharing of computer hardware.

Alternative concepts being investigated involve various combinations of centralized testing at some number of fixed sites (either at MEPS alone or some extension into high-testing volume locations) with computerized adaptive prescreening of applicants prior to CAT-ASVAB testing to minimize applicant travel.

Some of these concepts of operations have significant additional benefits that are hard to quantify. For example, a centralized system would eliminate remote and often unstandardized testing sites, thereby providing better control of the process and minimizing possible test take inequities. Under some scenarios, computer capability would be available for the recruiting services to use in processing of applicants. A final consideration is the need to look into the future, and develop a concept that can accommodate emerging testing technologies as they address military service requirements.

FUTURE TESTING SYSTEMS

What will be the new testing innovations after implementation of CAT-ASVAB? Two directions are already evident: (1) enhancements to the present aptitude tests in the battery; and (2) design of new types of measures that were infeasible to administer and score using test booklets (Wiskoff, 1986).

CAT-ASVAB Enhancements

Enhancements will come from opportunities presented by the operating characteristics of the adaptive testing methodology and from new developments in item response theory and CAT applications in this country and abroad. Developments are expected in the following areas:

Extraction of Additional Information from Ability Tests

The usual method of scoring a multiple-choice test is to treat each item independently and to make no allowances for how close the incorrect response is to the correct answer. Models are currently available to enable different approaches to the construction and scoring of ability tests, potentially resulting in improved measures. For those aptitudes where response time proves important, individual differences in speed and accuracy can be assessed and new scoring procedures developed. The extraction of additional information from ability tests can also be accomplished using new item formats providing more valid measurement than the multiple-choice format, possible in a shorter period of time.

New Techniques for Calibrating Test Items

Data collected for the purpose of calibrating new test items will be obtained "on-line," that is, by computer administration of a few items in addition to those determining the examinee's CAT-ASVAB scores. ONR and NPRDC are sponsoring a study to determine how best to implement the three-parameter logistic model, in estimating item and examinee parameters, to keep adaptive test scores comparable over a period of many years. New techniques are needed because, in adaptive testing, each examinee responds to only a small percentage of the test items comprising an aptitude area, *and these responses are scattered*, rather than occurring in blocks of fixed items as they would in paper and pencil testing. The new procedures under study are also expected to enable necessary adjustments in item difficulty, which reflect real changes in this item property over time.

Recent work by Mislevy (1986) on item calibration procedures has shown that the precision of estimates can be increased by taking advantage of dependencies between the latent proficiency variable and auxiliary examinee variables such as age, courses taken, and years of schooling. The incremental precision provided by auxiliary information is most useful in applications where few responses are obtained from each examinee, as in computerized adaptive testing.

New Adaptive Testing Methodologies

Promising work is underway on the development of what has been called a methodology for "second-generation adaptive testing" (Bejar, 1986). "First-generation adaptive testing" is characterized by its *declarative* nature, that is, an item pool is stored in a database along with item parameters. Second-generation adaptive testing instead follows a *procedural* approach: algorithms are constructed that generate the test items and control their psychometric characteristics. Thus, rather than calibrating specific test items, the *procedures that generate the items are calibrated*. Bejar (1986) points out that this next generation of adaptive testing has the following practical implications: (1) it is likely to be economical, in that the need to calibrate large numbers of items is eliminated; (2) the psychometric characteristics of the items constructing the test are better controlled; and, most importantly, (3) it is not sufficient to author, calibrate, and link items—it is also necessary to have a theory of item performance. The procedural approach to second-generation adaptive testing offers considerable promise for improving the validity of test-score interpretations, by continually submitting the theory of item performance to empirical testing.

Measuring New Domains

The content of ASVAB and consequently CAT-ASVAB has evolved from earlier forms of the paper and pencil battery, and from instruments like the service classification batteries. Presently, ASVAB includes those aptitude areas which have shown validity through prediction of training success in each of the services (United States Military Entrance Processing Command, 1984). For both CAT-ASVAB and ASVAB, new predictors of success in training or in "on-the-job" performance will be introduced as validity is established.

All branches of the military services are currently engaged in active research on the development of new predictors for potential inclusion in the selection and classification process. An ambitious ARI project, known as

"Project A," is underway with the following operational goals: (1) developing new measures of job performance which are useful criteria against which to validate selection/classification measures; (2) validating existing selection measures against both existing and project-developed criteria; (3) developing and validating new selection and classification measures; (4) developing a utility scale for different performance levels across military occupational specialties; and (5) estimating the relative effectiveness of alternative selection and classification procedures in terms of their validity and utility (Campbell, 1986). A trial predictor battery has been developed, including cognitive and noncognitive paper and pencil tests, and computer-administered tests of: reaction time, short-term memory, psychomotor precision, perceptual speed and accuracy, and other skills. Preliminary results indicate that the computerized measures show promise in predicting job performance in a diverse army sample, comprised of such occupations as infantry, cannon and tank crew, administration, and military policing.

AFHRL is conducting a multi-year basic research program on new asessment techniques known as the Learning Abilities Measurement Program (LAMP). The goals of this program are to: (1) specify the basic parameters of learning ability; (2) develop techniques for the assessment of individuals' knowledge and skill levels; and (3) explore the feasibility of a model-based system of psychological assessment (Kyllonen, 1986). As part of the LAMP program, computerized testing was used as a vehicle for exploring the dimensionality of information processing speed. In three separate studies, Air Force recruits were given a variety of computerized tasks designed to tap verbal, quantitative, reasoning, decision, classification, and choice skills (Kyllonen, 1985). Response time data was analyzed to determine whether a single speed factor could account for subject-to-subject variability, or whether multiple speed factors were required; in all three studies, a general speed factor was found, but the data could be better accounted for if separate factors were posited for reasoning, verbal, quantitative, perceptual processing, and memory search. Further research is considered necessary to develop a theory-based taxonomy of information processing speed variables from which assessment applications can be made.

NPRDC is exploring computerized tests of cognitive speed (Larson & Rimland, 1984) and spatial-visual ability (Hunt et al., 1987). A battery of cognitive speed tests measuring reaction time, in responding to simple and complex stimuli, and inspection time, in discriminating between two rapidly presented visual stimuli, were given to Navy recruits entering electronics school. Significant relationships were found for a male sample, between scores on these tests and rate of progress in training. This work was extended by exploring relationships between reaction time, auditory, and visual measures of processing speed and conventional cognitive tasks in a college sample (Saccuzzo, Larson, & Rimland, 1986). Evidence was found for a general mental speed factor, in addition to task specific sources of variability.

The Hunt et al. work has involved: (1) developing spatial-visual reasoning tests that take advantage of computer technology; and (2) determining if these tests measure dimensions of ability not measured by currently available tests. Thus far, 11 computer-administered tasks have been developed; six involve moving objects and five measure reaction time. The tasks were presented to college students along with eight conventional paper and pencil tests. Findings from this work are the following: (1) computer-controlled analogs can capture the individual variation in spatial-visual reasoning measured by conventional paper and pencil tests: (2) computer-controlled analogs are preferable because they provide measures of both speed and accuracy within a signle task and allow for the study of individual differences in speed accuracy trade-offs; and (3) the ability to work with moving elements in a spatial display is distinct from the ability to work with static visual displays. The authors state that, while the dynamic display tasks included in their study are presented for the first time, development of such tasks appears to be a promising area of psychometric research.

Additional work is needed before such promising experimental instruments can be used for personnel decision making; attempts to validate new measures with respect to training and job performance are underway in all services. Further study of diagnostic testing, where testing and instruction merge and examinees' knowledge bases and problem-solving skills can be measured as they learn new material, can be expected. The microcomputer is certain to serve as a powerful tool in much of future testing.

Cross-National Programs

ACAP is scheduled to be operationally tested and evaluated starting in October 1989, which means that for the first time there will be ASVAB scores of record for applicants which have been obtained in a computerized adaptive testing mode. Following the operational test and review of lessons leared from ACAP, full-scale development of a nationwide CAT-ASVAB will commence.

The investigation of CAT as a replacement for existing military test batteries has become a cooperative international pursuit. The Psychological Service of the German Federal Armed Forces is currently evaluating the use of computer-assisted and adaptive testing as part of an overall diagnostic system at recruitment centers (Steege, 1986). The experiment system employs IBM AT computers with 20mb hard disks for host and testing stations with the latter having specially designeed keyboard and headphones. The system, which has a local area network for up to 15 terminals, has been programmed with standard test batteries, apparatus tests, and adaptive instruments. Trials of the system have been underway for some time. The goal is to routinely apply CAT to the selection of all volunteers and draftees by 1988.

Early in 1987, the Belgian Army Forces introduced CAT testing as part of a two-day selection procedure for reserve officer candidates. The system consists of 24 linked test stations in two groups of 12. All tests are loaded on hard disk Wang computers with 10 megabyte memory. It is interesting to note that tests are administered in both French and Dutch to suit the bilingual population. Results from CAT and computer administered testing are provided immediately to interviewers for use in applicant processing.

The Israeli Defense Forces (IDF) have been evaluating various aspects of computerized test administration since 1982. Recent analyses on 320 soldiers evidenced comparable reliability, test intercorrelations, and item characteristics for paper and pencil and computerized tests (Dover, 1986). Soldier reactions were positive, only 5 percent reporting any difficulty in responding to the system. The IDF is planning to implement computerized testing in the near future and then address adaptive testing as the next systems improvement.

The considerable cross-national CAT technology transfer is evidenced by many technical exchanges, site visits, and data sharing. Recent outgrowths of this coordination have been the initiation of CAT research within the militaries of Australia, Canada, and Great Britain.

REFERENCES

Angoff, W. H. (1971). Scales, norms, and equivalent scores. In R. L. Thorndike (Ed.), *Educational measurement* (pp. 508–600). Washington, DC: American Council on Education.

American Psychological Association. (1985). *Standards for educational and psychological testing.* Washington, DC: Author.

_____ (1986). *Guidelines for computer-based tests and interpretations.* Washington, DC: Author.

ASVAB Working Group. (1980). *History of the Armed Services Vocational Aptitude Battery (ASVAB) 1974–1980.* Washington, DC: Office of the Assistant Secretary of Defense, Manpower, Reserve Affairs and Logistics.

Bejar, I. I. (1986). *Adaptive assessment of spatial abilities* (Contract No. N00014-83-C-0761). Arlington, VA: Office of Naval Research.

Bock, R. D., Gibbons, R., and Muracki, E. (1985). *Full-information item factor analysis* (MRC Report 85-1). Chicago IL: Methodology Research Center/NORC.

Campbell, J. P. (1986, August). *Project A: When the textbook goes operational.* Paper presented at the annual convention of the American Psychological Association, Washington, DC.

Day, L. E., Kieckhaefer, W. F., and Segall, D. O. (1986). *Predictive utility evaluation of computerized adaptive testing in the armed services* (Contract No. N66001-83-D-0343). San Diego, CA: Navy Personnel Research and Development Center.

Department of Defense. (1982). *Profile of American youth: 1980 nationwide administration of the Armed Services Vocational Aptitude Battery.* Washington, DC: Office of the Assistand Secretary of Defense, Manpower, Reserve Affairs and Logistics.

Dover, S. (1986, July). Introducing computerized testing to the organization—reactions of the organization and the subjects, and some test-retest data. In M. F. Wiskoff (Chair), *Computerized adaptive testing (CAT) applications: An international military perspective*. Symposium conducted at the 21st International Congress of Applied Psychology, Jerusalem, Israel.

Folchi, J. S. (1986, November). Communication of computerized adaptive testing results in support of ACAP. In W. A. Sands (Chair), *Computerized adaptive testing hardware/software development for the U. S. military*. Symposium conducted at the 28th annual conference of the Military Testing Association, Mystic, CT.

Gardner, E. F. (1966). The importance of reference groups in scaling procedure. In A. Anastasi (Ed.), *Testing problems in perspective* (pp. 172–280). Washington, DC: American Council on Education.

Greaud, V. A. and Green, B. F. (1986). Equivalence of conventional and computer presentation of speed tests. *Applied Psychological Measurement, 10,* 23–34.

Green, B. F., Bock, R. D., Humphreys, L. G., Linn, R. L., and Reckase, M. D. (1982). *Evaluation plan for the computerized adaptive vocational aptitude battery* (Research Report 82-1). Baltimore, MD: The Johns Hopkins University, Department of Psychology.

―――― (1984). Technical guidelines for assessing computerized adaptive tests. *Journal of Educational Measurement, 21,* 347–360.

Gullicksen, H. (1950). *Theory of mental tests*. New York: Wiley.

Hardwicke, S. B., and Yoes, M. E. (1984). *Attitudes and performance on computerized vs. paper-and-pencil tests*. Paper presented at the second annual Air Force conference on technology in training and education, Wichita Falls, TX.

Hunt, E., Pellegrino, J. W., Abate, R., Alderton, D. L., Farr, S.A., Frick, R. W., and McDonald, T. P. (1987). *Computer controlled testing of spatial–visual ability* (Report No. NPRDC-TR-87-31). San Diego, CA: Navy Personnel Research and Development Center.

Jones-James, G. (1986, November). Design and development of the ACAP test administration software. In W. A. Sands (Chair), *Computerized adaptive testing hardware/software development for the U. S. military*. Symposium conducted at the 28th annual conference of the Military Testing Association, Mystic, CN.

Joreskog, K., and Sorbom, D. (1986). *Lisrel VI: Analysis of linear structural relationships by the method of maximum likelihood, User's guide* (4th ed.). Mooresville, IN: Scientific Software.

Kieckhaefer, W. F., Segall, D. O., and Moreno, K. E. (1985). Medium of administration effects on attitudes toward ASVAB testing. *Proceedings of the 27th Annual Conference of the Military Testing Association, 1,* 55–60.

Kyllonen, P. C. (1985). *Dimensions of information processing speed* (Report No. AFHRL-TP-84-56). Brooks AFB, TX: Air Force Systems Command.

―――― (1986). *Theory-based cognitive assessment* (Report No. AFHRL-TP-85-30). Brooks AFB, TX: Air Force Systems Command.

Larson, G. E., and Rimland, B. (1984). *Cognitive speed and performance in basic electricity and electronics (BE&E) school* (Report No. NPRDC-TR-85-3). San Diego, CA: Navy Personnel Research and Development Center.

Lord, F. M. (1980). *Applications of item response theory to practical testing problems*. Hillsdale, NJ: Lawrence Erlbaum.

Martin, C. J., Park, R. K., and Borman, D. (1986, April). *Validating a computer adaptive testing system using structural analysis*. Paper presented at the annual meeting of the American Educational Research Association, San Francisco, CA.

McBride, J. R. (1982). *Computerized adaptive testing project: Objectives and requirements* (NPRDC Technical Note 82-22). San Diego, CA: Navy Personnel Research and Development Center.

Mislevy, R. J. (1986). *Exploiting auxiliary information about examinees in the estimation of item parameters* (Research Report No. RR-86-18-ONR). Princeton, NJ: Educational Testing Service.

Mitchell, P. A., Hardwicke, S. B., Segall, D. O., and Vicino, F. L. (1983). Computerized adaptive testing: A preliminary study of user acceptability. *Proceedings of the 25th Annual Conference of the Military Testing Association, 1*, 106-111.

Moreno, K. M. (1985, September). *Reliability and validity of CAT-ASVAB*. Paper presented at the meeting of the Defense Advisory Committee on Military Personnel Testing, San Diego, CA.

Nilles, J., Carlson, F. E., Gray, P., Holmen, M., and White, M. J. (1980). *Technology assessment of personal computers*. Los Angeles, CA: University of Southern California, Center for Future Research.

Owen, R. J. (1969). *A Bayesian approach to tailored testing* (Research Report 69-92). Princeton, NJ: Educational Testing Service.

Prestwood, J. S., Vale, C. D., Massey, R. H., and Welsh, J. R. (1985). *Armed Services Vocational Aptitude Battery: Development of an adaptive item pool* (Report No. AFHRL-TR-85-19). Brooks AFB, TX: Air Force Systems Command.

Rafacz, B. (1986, November). Development of the test administrator's station in support of ACAP. In W. A. Sands (Chair), *Computerized adaptive testing hardware/software development for the U.S. military*. Symposium conducted at the 28th annual conference of the Military Testing Association, Mystic, CN.

Saccuzzo, D. P., Larson, G. E., and Rimland, B. (1986). *Spped of information processing and individual differences in intelligence* (Report No. NPRDC-TR-86-23). San Diego, CA: Navy Personnel Research and Development Center.

Schmidt, F. L., Urry, V. W., and Gugel, J. F. (1978). Computer assisted tailored testing: Examinee reactions and evaluations. *Educational and Psychological Measurement, 38*, 265-273.

Schratz, M. K. (1985, September). *Psychometric and systems development features of the CAT-ASVAB Technical Manual*. Paper presented at the meeting of the Defense Advisory Committee on Military Personnel Testing, San Diego, CA.

———— (1986, August). *Test content considerations in the development of adaptive item pools*. Paper presented at the annual convention of the American Psychological Association, Washington, DC.

Steege, F. W. (1986, July). Computer assisted and adaptive testing as part of a system of measures of personnel psychology. In M. F. Wiskoff (Chair), *Computerized adaptive testing (CAT) applications: An international military perspective*. Symposium conducted at the 21st International Congress of Applied Psychology, Jerusalem, Israel.

Sympson, J. B. (1985, October). Controlling item exposure rate in computerized adaptive testing. In W. A. Sands (Chair), *Computerized adaptive testing research at the Navy Personnel Research and Development Center*. Symposium conducted at the 27th annual conference of the Military Testing Association, San Diego, CA.

Tiggle, R. B., and Rafacz, B. A. (1985). Evaluation of three local CAT-ASVAB network designs. *Proceedings of the 27th Annual Conference of the Military Testing Association, 1,* 23–28.

United States Military Entrance Processing Command. (1984). *Test manual for the Armed Services Vocational Aptitude Battery* (DoD 1304.12AA). N. Chicago, IL: Author.

Vale, C. D., Lind, J. E., and Maurelli, V. A. (1986). *Design of the CAT-ASVAB maintenance and psychometric facility* (Manpower and Personnel Laboratory Technical Note 86-7). San Diego, CA: Navy Personnel Research and Development Center.

Ward, W. C., Kline, R. G., and Flaugher, J. (1985, December). *College Board computerized placement tests: Summary of pilot testing results 1984–1985.* Paper presented at the ETS/DoD conference on testing technology and applications, Princeton, NJ.

Weiss, D.J., and Kingsbury, G. G. (1984). Application of computerized adaptive testing to educational problems. *Journal of Educational Measurement, 21,* 361–375.

Wharton, Y. L. (1977). *List of hypotheses advanced to explain the SAT decline.* New York: College Entrance Examination Board.

Wildgrube, W. (1985). News about CAT in the German Federal Armed Forces (GFAF). *Proceedings of the 27th Annual Conference of the Military Testing Association, 1,* 96–101.

Wiskoff, M. F. (1981). Computerized adaptive testing. *Proceedings of the National Security Industrial Association First Annual Conference on Personnel and Training Factors in System Effectiveness, 1,* 33–37.

—— (1986, December). *New directions in mental testing.* Paper prepared for Brookings Institution seminar, Washington, DC.

Wolfe, J. H. (1985). Speeded tests—Can computers improve measurement? *Proceedings of the 27th Annual Conference of the Military Testing Association, 1,* 49–54.

5   Evoked Brain Activity and Personnel Performance

Gregory W. Lewis
Richard C. Sorenson

INTRODUCTION

The primary thrust of the neuroscience research program at the Navy Personnel Research and Development Center (NPRDC) is to develop a technology for measuring evoked brain activity that correlates with the performance of military personnel and serves as a predictor of that performance. NPRDC has been investigating neuroelectric and neuromagnetic recordings to assess individual differences in brain processing and their relationship to on-job performance. Neuroelectric recordings include electroencephalography (EEG)

The opinions expressed in this chapter are those of the authors, are not official, and do not necessarily reflect the views of the Navy Department.

and the evoked potential (EP), while neuromagnetic recordings involve magnetoencephalography (MEG) and evoked fields (EF). The EEG (measured in microvolts) and MEG (measured in femtotesla) show minute ongoing brain activity. The EP and EF show the brain response averaged over several trials which result from precisely recorded visual or auditory stimulation. Neuroscience research has significantly advanced the understanding of brain function during the last several years. Such advances have been due to developments in electronics and computer technology as well as to progress in experimental procedures and data analytic tools.

It is our objective here to present a brief discussion of individual difference measurement and its history, and then to present evidence for the relationship between neuroelectric recordings and aptitude. We will emphasize the relationships between EP and on-job performance assessment that we have found over the last dozen or so years, and conclude with a brief discussion of new techniques that we and others are examining and developing to improve the sensitivity of brain function measures, using the neuromagnetic evoked field.

INDIVIDUAL DIFFERENCE MEASUREMENT

An Historic View of Psychological Testing

Dahlstrom (1985) provides an interesting review of developments in psychological testing that gives historical relevance to our current research in the use of neuroscience procedures for personnel assessment. He suggests that some of the earliest forms of testing were done by the Chinese between 2200 B.C. and 1905 A.D. The early Chinese testing covered not only writing ability in composing poetry and prose, but also the ability to reference important and wide-ranging topics of law, finance, agriculture, etc. Job sample tests were used to demonstrate abilities related to music, archery, horsemanship, and arithmetic (Buss & Poley, 1976). Dahlstrom claims the early personnel selection testing by the U.S., German, and British governments had roots in the early Chinese testing procedures. Other testing by Western governments, however, focused on more physical aspects of the subjects being tested, which included anatomy and physiology. Such focus can probably be traced to the Classicial Greeks, that is, Aristotle, Plato, Pythagoras, Hippocrates, and Galen. The Greeks suggested relationships between physiognomy (attributes related to physical features), anatomy, and physiology, and the person's character and ability. Dahlstrom states, "These authorities gave convincing rationales and explanations for individual differences in intellective insights, wisdom and judgment, emotional control, amiability and tolerance, as well as dispositions to depression, fearfulness, and psychosis" (p. 65).

Other relationship between body states and temperament and emotionality were suggested by de la Chambre (France) and Huarte (Spain) during the 1500s and 1600s. Dahlstrom (1985, p. 65) continues that Huarte devised an elaborate explanation of differences between individuals " . . . based upon hypothesized internal states like those employed by de la Chambre, as well as on body build and physiognomic features, and covered intellective, temperamental, and characterological attributes. Huarte not only provided specific means of character reading but prescribed methods for training and enhancing basic talents and capacities. This work met with great interest and wide acceptance, reflecting less the validity of his methods than the need for some means of dealing with and understanding human differences." Phrenology developed as a result of the interest in explaining behavior shifting from internal functions of the body to the brain. Even though phrenology proved ineffective as an assessment tool, it did provide a stimulus for further, more productive work in physiological psychology based on the conviction that the brain was the "organ of the mind," a phrase used by Gall, the German proponent of phrenology. Phrenology also contributed the important concept of localization of specific functions within the brain (Boring, 1950).

Measurement of mental abilities was first undertaken by Binet (France) and Galton (England), with Binet being the first to develop a practical psychometric instrument. Further refinement of ability testing came about when chronological age and mental age were used to compute a measure of intelligence (IQ = mental/chronological age), and through test standardization, statistical reliability/validity, and factor and other statistical analyses.

Testing of military personnel received impetus during World War I with the development of the Army Alpha test (verbal) and the Army Beta test (nonverbal). Again, Dahlstrom (1985) provides interesting reading concerning the development of the Army Alpha and Beta tests as well as the research performed on these test instruments (that is, effects of fatigue on test scores, interrelationships of subtests for various groups and individuals). Numerous tests were developed before, during, and after World War II by Yerkes, Thorndike, Cattell, Wechsler, Seashore, Strong, and others. College aptitude tests and the MMPI and other personality tests were developed along with tests designed for military purposes. The latter include the Armed Forces Qualification Test (AFQT), Armed Services Vocational Aptitude Battery (ASVAB), and others specific to the Navy. The Navy Basic Test Battery (BTB) subtests for military personnel assessment include the General Classification Test (GCT), the Arithmetic Reasoning Inventory (ARI), the Mechanical Aptitude Test (MECH), and the Electronics Technical Selection Test (ETST). The ASVAB replaced the BTB in 1976 for use in selecting and assigning Navy recruits.

Uses of Tests

There is now a multitude of test instruments and procedures. *The Ninth Mental Measurements Yearbook* lists in excess of 1,400 tests that provide scores on over 6,000 variables (Mitchell, 1985).

Psychological measurement instruments can be clustered in terms of the degree to which responses to items can be judged correct or incorrect. If there is no correct response, the test may be considered to be noncognitive. Noncognitive tests include those that measure personality, interests, values, beliefs, and attitudes. In short, the noncognitive tests tap the subjects' model of the real world and how they feel about elements therein. If the responses can be judged "right" or "wrong," the test is spoken of as a cognitive test. Cognitive tests measure achievement, abilities, and aptitudes. The items in the various types of cognitive tests may be very similar; the distinction is made in terms of the use of the test.

Achievement tests measure *past* performance—particularly learning performance. Achievement measures are valid in the sense of content validity. A particular domain of skills and knowledge must be identified and items prepared which sample that domain. The achievement measure is valid to the extent that the items are properly prepared and the domain is adequately sampled, usually as judged by subject matter experts.

Ability tests usually measure *present* performance. They are properly validated in the sense of construct validity. A measure of a particular ability is correlated with other ability measures. If the measure is highly correlated with other measures of the same ability and uncorrelated with measures of other abilities, then one can conclude that the ability measure is valid.

Aptitude tests are related to *future* performance. They are used to predict future job and educational performance, and are validated in terms of criterion validation—either predictive or concurrent. An aptitude test is considered valid to the extent its scores relate to criterion behavior, such as performance in a particular course or on a specific job task. Ideally the test is administered to a representative sample of those to be considered for selection (that is, predictive validation). A less advisable approach is to study the relationship between the predicting test and criterion behavior in those already selected and performing the activity (that is, concurrent validation). The problem with concurrent validation is that the sample does not represent a group from which selection will eventually be made—selection by some means has already occurred—and the experience of performing the criterion behavior has conceivably changed the individuals. The individuals perform differently on the test because of the experiences of course instruction or task performance. Traditionally, the degree of validity is expressed in terms of correlation coefficients between the aptitude measure and the criterion measure. Contingency tables showing success rate or average performance for differing intervals of the aptitude score are generally more informative.

Most cognitive measurement techniques depend on an evaluation of the products of cognition. We compare the answers to questions against a standard. An alternative is to measure the cognitive *processes* rather than their *products*. For instance, in the past, attention has been given to the measurement of eye movements (Dillon & Wisher, 1981), reaction latency (Jensen, 1985; Larson & Saccuzzo, 1986), and inspection time (Saccuzzo, Larson & Rimland, 1986). The approach we are reporting on, however, can be characterized as process measurement of aptitudes and involves both predictive and concurrent validation.

Individuals differ in their ability to adapt to training and job demands. Such adaptability depends on factors such as basic aptitude and the ability to tolerate new and stressful situations.

Traditional personnel assessment in both military and civilian communities has depended heavily on paper and pencil tests that predict school and training performance fairly well, but do not predict on-job performance with the same degree of accuracy. Linn (1982, after Ghiselli, 1966) examined relationships between aptitude test scores and training and proficiency criteria. Validity coefficients averaged between 0.30 and 0.40 for training criteria and less than 0.20 for proficiency criteria. For example, with training criteria, they found validity coefficients of 0.47 for clerks, 0.35 for protective services (fire, police), 0.54 for personal services (hospital attendants), 0.15 for vehicle operators, 0.41 for trades and crafts, and 0.40 for industrial workers. Validity coefficients of aptitude test scores with proficiency criteria for these same groups were 0.27, 0.23, 0.03, 0.14, 0.19, and 0.16.

An extensive literature review on predicting military job performance was published by Vineberg and Joyner (1982). This review covered the years 1952–82. They found aptitude variable correlations of about 0.40 for job knowledge, 0.10 to 0.35 for job sample tests, 0.24 for composite measures of suitability, and 0.15 for global ratings of performance. Hunter and Hunter (1984) found that cognitive tests correlated with on-the-job performance in the low to moderate range depending on the complexity of the information processing requirements of the job. For certain highly skilled people, such as aviators and sonar operators, training attrition is still too high. We must explore new techniques for predicting performance and selecting personnel.

The introduction of computer adaptive testing (CAT), based on item response theory and adaptive testing algorithms, has allowed more reliable measurement of personnel at all ability levels. Although CAT provides the potential for some improvement in personnel assessment, it remains to be demonstrated whether CAT will be able to adequately predict on-job performance. Its impact is likely to be more in the realm of spatial-visual reasoning, memory, and attention than in the measuring of verbal intelligence or psychomotor abilities (Hunt & Pellegrino, 1985). The applications of CAT in large-scale selection testing are currently being explored at

NPRDC. Other new approaches involve the measurement of cognitive processes rather than the products of cognition.

To improve attitude measurement and prediction of on-job performance and attrition, new testing procedures are required to supplement the information from existing tests. Neuroscience produces (for example, neuroelectric and neuromagnetic recordings) that measure brain processes have shown promise for improved performance prediction (Lewis, 1983a; Lewis et al., 1986). Researchers have demonstrated that such procedures generate reliable measures useful in discriminating normal populations from populations with cognitive dysfunctions and disorders (John et al., 1983). It is hoped that assessing brain information will facilitate more accurate prediction of performance under training, nonstressed (baseline), fatigue, and stressful on-job performance conditions.

NEUROELECTRIC MEASURES OF BRAIN ACTIVITY

EEG/EP Individuality and Test-Retest Reliability

Neuroelectric or neuromagnetic measures for personnel assessment must be sensitive to individual differences and show long-term stability or reliability. Stability refers to the similarity in the waveforms of an individual across time. Stability over time is a preprequisite for using neuroelectric data for personnel assessment and has been demonstrated in this and other laboratories (Lewis, 1984). The usual procedure for determining stability involves computing the correlation coefficient between two waveforms (Glaser & Ruchkin, 1976). High correlation suggests stability, while low correlation suggests variability.

Early papers by Travis and Gottlober (1936, 1937), Davis and Davis (1936), Rubin (1938), and Williams (1939) suggested that EEG activity showed individuality and was stable from day to day. Such activity patterns were shown to be not only stable and individualistic, but also inherited (Lennox, Gibbs, & Gibbs, 1945). More recently, several studies have found stable EEG records within subjects (Fein et al., 1983; Matousek, Arvidsson, & Friberg, 1979; Stassen, 1980; Van Dis et al., 1979). Research has also demonstrated that EEG and stimulus-locked EEG records (EP) are very sensitive to individual differences (Berkhout & Walter, 1968; Brazier, 1962; Buchsbaum & Pfefferbaum, 1971; Callaway, 1975; Henry 1941a,b; Lewis, 1983a; Uttal & Cook, 1964; Werre & Smith, 1964). Early research relating more variable, statelike attributes (for example, anxiety arousal) to psychological aspects were described by Travis (1937), Travis and Egan (1938), Hoagland, Cameron, Rubin, and Tegelberg (1938), Knott (1938), and Hadley (1940, 1941).

With the advent of improved instrumentation and signal averaging techniques, tighter stimulus-response observations have become possible in neuroscience research. Sensory systems (that is, visual, auditory, somatosensory) as well as higher order cognitive processing and psychological variables can now be explored in greater detail and with greater precision. However, much variability in EP recordings has been noted between and within subjects. EP variability, its contributing factors, and its relationship to cognitive variability have been discussed by Callaway (1975). Greater EP variability has been noted in patients with mental and behavioral disorders (Buchsbaum & Coppola, 1977; Callaway, 1975; Callaway, Jones, & Donchin, 1970; Cohen, 1972; Shagass, 1972) and in newborn infants (Ellingson, 1970), but not in normal adults. Dustman and Beck (1969) reported the stabilizing of visual EP amplitude with maturity at about age 16. Ellingson, Lathrop, Danahy, and Nelson (1973), studying adults and infants, found greater visual EP stability within sessions than over days; and adults showed greater stability than did the infants. These authors used the Pearson product-moment correlation on the 500 msec visual EP (128 data points) for their stability measure.

Test-retest correlations for the EP have been reported to range from about 0.70 to 0.90 for varying modalities (visual, auditory, somatosensory), subject age groups, and measures (amplitude, latency, slopes). Stability of evoked activity has been examined for the visual modality (Dustman & Beck, 1963; Kooi & Bagchi, 1964; Wicke, Donchin, & Lindsley, 1964); the auditory modality (Buchsbaum, Henkin, & Christiansen, 1974; Ellingson, Danahy, Nelson, & Lathrop, 1974); and by comparing visual and auditory modalities (Buchsbaum & Coppola, 1977). Results have shown high within-subject and low between-subject stability. For the visual modality, greatest stability has been found in the occipital and central regions (Kooi & Bagchi, 1964). Auditory EP stability has been greatest for children (six to nine years) and least for older adults (40 to 60 years) (Buchsbaum et al., 1974). Comparison of visual and auditory recors showed grater stability for the visual than for the auditory modality (Buchsbaum & Coppola, 1977). Their area-under-curve measures showed greater stability than baseline-to-peak measures for records obtained two or more weeks apart.

In a recent study (Lewis, 1984), visual, auditory, and bimodal (visual plus auditory) EP records were obtained about two hours apart from a group (*N* = 8) of young adult males. Their ages averaged 19.6 + / − 0.9 years (range: 19–21). Similar records were obtained about two months apart from a group of older adults (*N* = 7 males, 1 female). Ages for this group averaged 33.1 + / − 9.6 years (range: 21–43). Waveform stability was assessed using a cross-correlation measure, similar to that used by Glaser and Ruchkin (1976). No statistically significant EP amplitude or temporal stability differences were found between the two groups. Age, however, was positively correlated with visual EP stability measures in the occipital area, and negatively correlated with auditory stability measures in the temporal and parietal areas. No correlation of age was found with the bimodal stability measures. Large subject differences were found for EP analog waveform amplitude and temporal stability. The EPs were highly stable within subject from session to session, whether they were recorded hours or months apart. Differences in patterns of waveform stability existed for site and modality conditions across individuals. The degree of intra-subject waveform stability may be considered a personnel assessment measure and has been shown to be related to on-job performance. We will discuss this in greater detail later. Greatest stability was found for the bimodal presentation ($r = 0.70$-0.90), less for the visual records ($r = 0.60$), and least for the auditory records ($r = 0.50$). For visual stimuli, mean correlations were 0.70-0.80 in the occipital area, decreasing to about 0.40 in the frontal area. Auditory reliabilities were highest in the frontal/temporal area (0.60), while those for bimodal stimuli were greatest in the parietal/occipital area (0.70-0.90).

Sensory interaction and integration of the visual and auditory modalities appear essential for adequate performance of complex tasks such as reading (Lewis & Froning, 1981; Shipley, 1980). Bimodal records often produce greater amplitude and shorter latency of EP components than visual or auditory records alone, suggesting sensory integration. Integration of the two sensory systems is probably the main contributor to the greater stability of bimodal records compared to that of visual or auditory records taken separately. Data presented in the Lewis (1984) article suggested that bimodal presentation may activate greater populations of brain fibers, a quantitative factor contributing to waveform stability.

Brain Activity and Ability

The first reporting of human brain activity was by Berger in 1929. Relationships between EEG records and test intelligence date back more than 50 years (Berger, 1933 [cited in Vogel & Broverman, 1964]; Kreezer, 1937, 1938, 1939; Kreezer & Smith, 1936, 1937) and dealt with psychopathic personalities and those with lower mental ability. Vogel and Broverman (1964) have reviewed the literature and assessed the positive and negative research results. Their references included 68 citations, most of which were EEG/IQ studies. They suggested that the most consistent EEG/IQ relationships were found for children, for individuals with brain injuries, and for patients with very low mental ability or who have been institutionalized for other reasons. Little relationship was found for normal adult subjects. Vogel and Broverman also pointed out that those researchers who showed negative results often limited their recording sites to the occipital region to obtain occipital alpha (8-13 Hz). Those studies that found EEG/IQ relationships generally used other recording areas (that is, frontal and parietal). They also suggested that the EEG measures in children were probably related more to absolute mental ability than to IQ. Large differences between subjects in the EEG and EP may be due to differences in age, with very young and old subjects showing longer latencies and larger amplitudes than young adults (Callaway, 1975). They commented on several methodological problems that may have weighed against finding more solid EEG/IQ relationships. These included (1) the measurement of intelligence, which was often confounded with age; (2) EEG recording sites (in areas other than the occipital region); (3) conditions during recording, including the fact that nearly all studies recorded the EEG while subjects were idle and not performing mental tasks; (4) use of subjects of both sexes in the smaples, even though large sex differences are reflected in EEG measures; and (5) restricting the EEG measures to the traditional frequency bands (that is, delta, theta, alpha, beta). Nevertheless, higher frequencies in varying brain regions and the absence of slower delta and theta rhythms were found to be associated with higher levels of intelligence.

One of the earliest EP studies dealing with intelligence was published more than 20 years ago (Chalk & Ertl, 1965), followed by several other papers (Ertl, 1968, 1969, 1971, 1973; Ertl & Schafer, 1969). Ertl proposed the idea that "neural efficiency" is related to IQ, that is, speed of information processing is related to IQ. Smart subjects would have shorter visual EP latency components than less smart subjects. Ertl later developed and sold the neural efficiency analyzer, which caused much interest and controversy and stimulated a popularization of the approach (Helvey, 1975). Besides the fact that he used unconventional recording locations, Ertl did not use a conventional latency measure. Callaway (1975) covers the relationships between EP latency and intelligence in considerable detail, including the impact that Ertl had in this area of research.

Ther have been several replications of the Ertl work, including that by Shucard and Horn (1972), Galbraith, Gliddon, and Busk (1970), and Callaway (1975), the latter using Navy recruits. Callaway's own work was reported at length in his book.

Frequency and latency are inversely related, shorter latency being associated with higher frequency. If shorter latency is related to high IQ, then higher frequencies are also related to high IQ. Bennett (1968) studied the relationships between the dominant frequency in the visual EPs of 36 subjects and their IQ scores, as measured by the Wechsler Adult Intelligence Scale. He found a statistically significant correlation of 0.59. Even though he did not report the probability level, using 35 df, the $r = 0.59$ value does exceed the $\alpha = .01$ significance level. Weinberg (1969) studied 42 subjects who had IQ scores ranging from 77 to 146, measured by the verbal portion of the Wechsler Adult Intelligence Scale. EP data were obtained by having the subjects passively observe a visual stimulus. He found that the 12 and 14 Hz frequency components showed statistically significant correlations with IQ test scores.

Ertl (1971) attempted to replicate both the Bennett and Weinberg studies, but was unable to do so. He did suggest that higher IQ subjects tended to show higher frequency components during the first 200 ms of the EP than did the lower IQ subjects. From 200-500 ms, both IQ groups showed about the same frequency component amplitudes. Ertl (1973) found EP/IQ relationships by examining 80-ms windows within the 0-240-ms portion of the EP waveform. Shucard and Callaway (1973) were not able to find statistically significant relationships between EP frequency components and IQ.

Everhart, China, and Auger (1974) tested Ertl's neural efficiency analyzer (NEA) to see if it acutally measured visual EPs or not. They found no difference between experimental conditions for the visual stimulus, regardless of whether the stimulus was "on" or "off" or whether the presentation of an auditory stimulus was "on" or "off." They concluded that the NEA was measuring relationships between ongoing EEGs, not EPs, and verbal intelligence. In addition, they suggested that the NEA not be used to assess or predict verbal intelligence because the correlations were too small to be of value.

Major negative findings concerning the NEA were reported by Davis (1971). Virtually no relationships were found ($r = 0.0 +/- .15$). The Davis study, according to Callaway (1975), provided little resolve of the issue due to claims of very noisy data and lack of oversight. Lykken (1973) severely criticized this study and stated that "this 'replication' was a debacle, an enormously expensive, total failure" (p. 463). Rhodes, Dustman, and Beck (1969) and Dustman and Beck (1972) were also unable to replicate EP latency/IQ relationship.

Other aspects of the EP/IQ relationship that have been investigated include recovery functions, which were generally slower for slow learners than for college students (Wasman & Gluck, 1975), and number and amplitude of potentials during conditioning in children, which were found to be weaker for lower IQ subjects (Lelord, Laffont, & Jusseaume, 1976).

In addition to the speed-of-processing theory of intelligence, Hendrickson and Hendrickson (1980) suggested that intelligence is related to error rates in the brain. They proposed that the way information is coded and transmitted within the brain determines the error rates during cognitive processing. HIGH intelligence results from low error rates within the brain. With increased error rate, there is lower EP amplitude and less complexity in the EP waveform components. The "string measure" was used to measure complexity and amplitude by literally laying a string on the waveform and measuring the length. The greater the amplitude and complexity, the longer the string length. Eysenck and Barrett (1985) reviewed at considerable length this error rate theory, as well as other proposed interactions of psychophysiology and intelligence. Blinkhorn and Hendrickson (1982) showed very strong relationships between the string measure and intelligence, as measured by the Raven's Advanced Progressive Matrixes (APM). They presented auditory tones to 33 university students and found a correlation of 0.54 between the auditory EP string measure and AP, score ($p < .001$).

Vetterli and Furedy (1985) took issue with the string measure used by the Hendricksons based on empirical tests of error theory and speed theory hypotheses.[1] Vetterli and Furedy state that the Hendrickson string measure was not only arbitrary, but the correlations that were found depended on the way in which the EP amplitude and time axes were plotted. They suggested that the greater the ratio of the amplitude to time, the higher the correlation between EP and IQ. This string measure had been revised by the Hendricksons in an attempt to eliminate the arbitrary nature of their measure. Vetterli and Furedy compared this revised string measure with a latency measure and an average voltage measure to assess EP complexity. They used two data sets, one from select subjects used by Ertl and Schafer (1969), and a second smaller data set from Weinberg (1969). Their results tended to support the speed theory rather than the error theory with regard to EP/IQ relationships.

In our early work (Lewis, Rimland, & Callaway, 1977), we used procedures similar to those of Ertl and Schafer (1969) in order to examine the relationship of the neural efficiency measures to aptitude. In our work, the visual stimulus was triggered by the subject's own background EEG activity. The reliability of the EP latency measures was increased by taking into account each subject's background EEG activity. EPs typically crossed the baseline several times within 500-600 msec. We obtained latency measures to the first, second, and third positive-going crossings (that is, positive-slope zero-cross). Our subjects were 206 Navy recruits who scored high (*n* = 103) and low (*n* = 103) on the Armed Forces Qualifications Test (AFQT). The HIGH group scored between the 80-99 centiles on the AFQT, which corresponded to an IQ range of 113-133. The LOW group ranged between 20-40 centiles on the AFQT, which corresponded to an IQ range of about 87-96. Even though we did not find statistically significant differences between our two groups, the latency values tended to generally follow the expected directions (that is, HIGH group had shorter latencies than did the LOW group—HIGHs = 89, 185, 283 ms; LOWs = 92, 189, 290 ms). Also, we did not find statistically significant differences (biserial correlation) between latency measures and criterion grouping of remedial readers (that is, PASS versus FAIL) (Lewis, Rimland, & Callaway, 1977).

MODELS OF BRAIN PROCESSING

Most of our research has emphasized EP amplitude measures, as they appear to be the most appropriate measures for assessing the models of brain processing that we have been following in the literature. These models include lateral asymmetry, variability (temporal and spatial), and resource allocation. Our work will be discussed within the context of one or more of these theories.

Amplitude

The amplitude measure of early choice was the microvolt root mean square (uVrms). There are several advantages to using the rms measure: It is easily computed, is objective and not dependent on visual inspection and identification of EP components, and has a high correlation with low frequency components, of which the averaged waveforms are primarily composed. Two disadvantages are associated with the use of the rms measure: It does not assess latency, nor does it retain the polarity information of the signal. The rms is computed over a time interval. As such, latency can be fairly well-estimated by narrowing the time interval or window. We have found the uVrms to be effective in assessing individual differences from a large and varied group of subjects, because not all subjects show well-defined EP components (Lewis & Froning, 1981).

In one of our early reports, we examined the relationships between amplitude measures and reading ability (Lewis, Rimland, & Callaway, 1976). The subjects ($n = 73$) included those from a high-risk group with a greater than average probability of early discharge from the Navy. All subjects were male recruits, with an average age of 19 years, who had been admitted into the Navy despite a rather poor level of reading ability. They scored between the twentieth and fortieth centiles on the AFQT. In addition, these recruits scored between 3.0 and 5.5 grade levels on the Gates-MacGinitie Reading Test, meaning that they could not read as well as the average 11- or 12-year-old. Some of the recruits ($n = 32$) improved their reading ability sufficiently to continue on active duty (ACT group), while 41 failed remedial reading training and were discharged from the Navy (DIS group). The AFQT scores for the groups were 35.7 $+/-$ 7.8 centiles for the ACT group and 33.7 $+/-$ 10.0 centiles for the DIS group. Entering reading grade levels were 4.4 $+/-$ 0.7 (ACT) and 3.9 $+/-$ 0.7 (DIS).

A statistically significant biserial correlation ($r = 0.32$, $p < .05$) was found between EP amplitude (rms) at the F4 site (Jasper, 1958) and the group criterion (ACT/DIS). A discriminant analysis was performed for the two groups that found statistical differences between them ($F = 5.59$, $p < 0.25$). Cross-validation was obtained using the Training-Test Set procedure ($X^2 = 5.56$, $p < 0.25$). No group differences were found using the trial-to-trial variability or latency measures. Amplitude measures were greater for the ACT group than for the DIS group, which may reflect greater temporal variability for the DIS than for the ACT group. Greater variability, or "jitter," in the single epochs is generally reflected as lower waveform amplitudes in the EP average.

Follow-up performance records were obtained for enlisted recruits three years after recording the initial EP data. The primary objective of recording the original EP data was to compare the EP amplitude and asymmetry predictors with the traditional paper and pencil aptitude and acedmic predictors used by the Navy. The subjects were the same used in earlier projects (Lewis, Rimway, & Callaway, 1976, 1977). Not all subjects' L records were available for the follow-up research. The sample ($N = 173$) was divided into two groups based on the number of promotions each enlistee achieved during the preceding three years. The HIGH group ($n = 102$) had two or more promotions, while the LOW group ($n = 71$) had fewer than two promotions.

Table 5.1 shows both groups to be fairly similar with respect to GCT scores and highest level of education (HIED) reached. Both groups averaged a

Table 5.1
Reading Grade Level, Aptitude Test Scores, and High Education Level Achieved for the HIGH and LOW Promotion Groups

	HIGH		LOW	
	MN	SD	MN	SD
RGL	9.48	2.58	8.93	3.27
GCT	52.53	13.28	52.82	13.88
AFQT	57.75	30.65	61.07	29.01
HIED	12.01	.76	12.06	1.19

twelfth-grade (high school) level of education. The HIGH group averaged higher in reading grade level RGL (9.5) than did the LOW group (8.9). However, the LOW group scored higher on the AFQT (61 centiles) than did the HIGH group (58 centiles). The mean standardized score for the AFQT is $50 +/- 10$ centiles, which suggests that the LOW group averaged one (1) standard deviation above the mean on the AFQT. However, this observation must be considered in light of the fact that both group standard deviations were large (and similar).

Sixteen EP amplitude variables (two series of 50 flashes each for the eight sites) and four aptitude-academic (traditional paper and pencil) variables served as input to a discriminant analysis (DA). The four paper and pencil test variables included scores from an aptitude test (AFQT), a classification test (GCT), reading grade level of the Gates-MacGinitie Reading Test (RGL), and the highest grade of education completed (HIED). Five EP amplitude measures (F4, C3, O1, O2, P4) differentiated the two groups ($F = 4.15, p < 0.01$) into either the HIGH or LOW promotion group more effectively than did the traditional paper and pencil predictors. The hold-out sample procedure (training set, test set) was used for cross-validation and was statistically significant ($X^2 = 9.78, p < .005$) with 62 percent of the test set being correctly classified. RGL and AFQT entered the DA at steps 6 and 7 but did not enhance the cross-validation result ($X^2 = 5.03, p < .025$). Correctly classified cross-validation cases dropped to 58 percent.

Lateral Asymmetry

The lateral asymmetry model has received much professional and popular press attention (Buchsbaum & Fedio, 1969; Dimond & Beaumont, 1974; Galin & Ellis, 1975; Kinsbourne, 1972, 1978; Knights & Bakker, 1976; Mintzberg, 1976; Ornstein, 1977, 1978), and critical review (Beaumont, Young, & McManus, 1984). This model suggests that logical, sequential, and analytic processes are performed in the left hemisphere, while spatial, simultaneous, and integrative processes are performed in the right hemisphere.

Asymmetry may be measured as the EP amplitude difference between the left (LH) and right (RH) hemisphere (RH minus LH) for homologous sites. We felt that most paper and pencil tests and classroom instruction would primarily assess functions attributed to LH, and that such functions might be assessed through the lateral asymmetry model (Lewis & Rimland, 1979). Many, if not most, Navy on-job and other real-world performance would depend also on the functions usually attributed to RH, for example, requiring integrative, spatial, and judgmental skills. We found few, if any, relationships between lateral asymmetry measures and academic criteria (for example, remedial reader test scores, (Lewis, Rimland, & Callaway, 1976)) or aptitude (AFQT) (Lewis, Rimland, & Callaway, 1977).

In order to assess this model, we used groups of highly skilled personnel. EP data were recorded from 26 sonar operator trainees (Lewis & Rimland, 1980) and 58 aviators (28 pilot, 30 radar intercept officers (RIOs)) in an operational environment (Lewis & Rimland, 1979). The operator of today's sophisticated sonar equipment must perform difficult and demanding mental operations requiring quick processing of visual and auditory information and the visualization of moving objects in three-dimensional space. Although conventional paper and pencil aptitude tests were reasonably effective in predicting academic performance in sonar school, they are not effective in identifying, from a pool of applicants, those who are most likely to perform successfully as sonar operators.

Performance measures were obtained from the 26 trainees, which included instructor and peer ratings, aptitude test scores, classroom grades, and laboratory test scores. The laboratory performance scores were based on electronic test equipment operation, aural identification of sonar contacts, visual identification of sonar contacts, and sonar simulator performance. Two groups were formed based on the sonar simulator performance scores. Scores on the aptitude tests and laboratory tests were very similar for the two groups (Table 5.2). The fourth lab test (Sonar Simulator Performance) was used as the criterion. Correlation of the performance score with instructor ratings was significant ($r = -.74, - < .01$), as was the correlation with peer ratings ($r = -.60, p < .01$). Negative correlations were due to the scoring of rating questions. Instructor and peer ratings agreed well ($r = .66$, $p < .01$). The magnitude of these correlations ($r = .66, p < .01$). The magnitude of these correlations ($r = .60-.74$) attests to the reliability of the measures. Performance score also agreed with overall classroom grade ($r = .35, p < .01$). The AFQT score predicted the classroom grade ($r = .47, p < .01$), but not the simulator performance score ($r = .06$, NS).

If good simulator performance depended on RH functioning, the high performers should show larger RH amplitude and, therefore, positive asymmetry (that is, RH > LH). Group differences were found primarily in the occipital region ($F = 5.87, p < .025$). The data showed that the high

Table 5.2
Descriptive Statistics for HIGH and LOW Sonar Trainee Performance Groups

Item	HIGH (n = 14)		LOW (n = 12)	
	X̄	SD	X̄	SD
Aptitude Test Scores				
GCT	60.46	6.91	60.33	5.60
ARI	57.23	6.47	57.92	6.37
MECH	54.54	6.98	52.00	6.62
ETST	64.91	4.28	63.25	2.73
AFQT	73.71	13.05	72.33	12.07
Test Scores on Sonar Laboratory Practicals:				
Elect. Test Equip. Oper.	89.25	8.51	86.08	9.39
Aural Ident. of Sonar Contracts	89.36	9.25	88.92	11.07
Visual Ident. of Sonar Contracts	77.17	17.84	73.42	17.45
Sonar Simulator Performance	87.57	4.31	74.92	7.79
Sonar Classroom Grade	75.43	4.88	72.75	5.71
Age	20.54	2.25	19.22	.82

performers had positive asymmetry (0.18 uVrms) in this region, while the low performers had large negative asymmetry (-0.52 uVrms). These asymmetry values showed a difference that was statistically significant ($t = 2.70$, $df = 18$, $p < .02$). The standard deviations for these asymmetry measures were much larger for the low performers (0.75 uVrms) than for the high performers (0.35 uVrms). These findings are also consistent with other research showing the right hemisphere to be heavily involved in tasks relating to the individual's orientation in three-dimensional space. The finding of large differences between the high and low group in recordings taken from the occipital area is of special interest because of the occipital area's strong role in visual perception. The operational tests used in sonar student selection did not distinguish between the high and low groups (Lewis & Rimland, 1980).

Pilots and RIOs might be considered to represent prototypes of the two different kinds of information processing served by the right and left hemispheres, respectively. Pilots must be able to cope quickly with problems in three-dimensional space and to make correct split-second judgments based on incomplete information (presumably nondominant RH functions). While RIOs must perform many pilotlike tasks, many of their duties required them to deal with explitic information in a sequential, logical, and systematic way (alleged dominant LH functions). Obviously, pilots must also have LH abilities, and RIOs cannot succeeed without RH spatial and judgmental abilities. Careful screening of aviation candidates ensures that both pilots and RIOs have above-average intellectual abilities, particularly the more readily measurable LH skills. Nevertheless, the key elements of pilot and RIO performance might be reasonably categorized as primarily right- and left-hemispheric in nature, respectively. This reasoning leads to the hypothesis that the pilot group may be discriminated from the RIO group based on visual EP amplitude measures from the left hemisphere and right hemisphere. This assumes that classification of aviation officers into the pilot and RIO groups, and/or subsequent on-job experience in these respective groups, may lead to a differentiation of the two groups in terms of brain functioning.

A second hypothesis consistent with this reasoning is that the quality of pilot and RIO performance may be a function of the degree to which the pilots possess relatively superior RH abilities and RIOs possess relatively superior LH abilities.

Pilots and RIOs may, in fact, represent prototype groups for hemisphere asymmetry research. Doktor and Bloom 1977) used electrophysiological methods in a similar way in their study of operations researchers (LH) and holistically-oriented company executives (RH). They reported a change from right to left hemisphere of EEG alpha activity (in the temporal area) for the operations researchers when they performed spatial as opposed to verbal tasks. No differences were found for the executive group. Lawyers (primarily verbal, LH) and ceramists (primarily spatial, RH) were used in the Galin and Ornstein (1974) study of hemipheric functioning in contrasting occupational groups. Galin and Ornstein used Kinsbourne's (1972) technique of measuring gaze shift of eyes during thought and in response to questions as indicators of hemispheric activation. Although they found gaze shift differences between lawyers and ceramists, no EEG differences were found between these two groups.

Since the pilots and RIOs performed somewhat different functions, we wished to find if they differed, as groups, in their visual EPs. Any difference found might be due to either selection factors or experience. Initial self-selection to be a pilot or an RIO or explicit selection, as well as self- or operational selection during or after training, could result in groups of pilots or RIOs that were quite dissimilar. It might also be argued that different experiences during or after training could cause EP differences.

Both uVrms mean and standard deviation values of the pilots exceeded those of the RIOs. Discriminant analysis was used to estimate the extent to which the visual EP amplitude measures might differentiate the pilots from the RIOs. EP amplitude measures at the C3 and F3 sites discriminated ($F = 6.53, df = 1, 56, p < .025$) and correctly classified 71 percent of the aviators.

In addition to our interest in possible group differences between pilots and RIOs, we wished to determine if *individual* differences in performance within the pilot and RIO groups might be reflected in EP measures.

It had hypothesized that high-rated pilots would show greater activity than low-rated pilots, and that high-rated RIOs would exhibit greater LH activity. Each of the 28 pilots and 30 RIOs was compared to the other pilots and RIOs, respectively, on a scale of 1 to 10. Ratings were obtained from their operations officer, a former Navy Blue Angel aviator qualified to assess flying proficiency as well as ground school performance. To test these hypotheses, asymmetry values were computed from homologous sites in the frontal, central, parietal, and occipital regions (that is, RH minus LH in each of the four regions) and plotted against the performance ratings for the pilot and RIO groups. The percentage of aviators rated 8, 9, and 10 whose RH amplitude was greater than their LH amplitude was then determined. The clearest asymmetry relationship was seen in recordings from the parietal region (Figure 5.1).

Figure 5.1
Percent of Right-Handed Aviators with Parietal RH Amplitude Greater than LH Who Were Rated 8, 9, or 10 on a Scale of 1 to 10 by Their Operations Officer (from Lewis & Rimland, 1979)

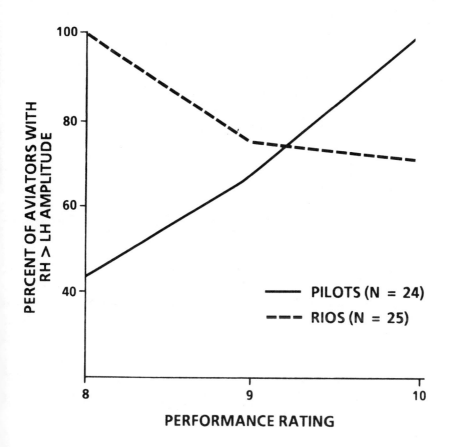

Several limitations of this study should be noted. First, the samples were composed of experienced aviators; thus, the results were confounded by restriction of range and the effects of experience. Second, the number of subjects tested was too small to permit cross-validation of findings. Third, pilot and RIO performances were rated by only one person and were of limited range and reliability. Ideally, objective performance and/or simulator-derived proficiency measures should be used. Finally, the stimulus used to evoke the brain potentials consisted of a simple flashing light. No dynamic task was performed.

Despite the limitations of this study, several promising findings were observed. These included EP differences between pilots and RIOs that provided preliminary confirmation of the hypothesized differences between right hemisphere and left hemisphere functioning in pilots and RIOs.

In discussing the EP differences between pilots and RIOs, we noted the possibility that the EP findings could be the result of endogenous and/or experiental factors. Subsequent analyses relating flight experience to the EP measures showed markedly greater EP asymmetry dispersion among aviators (pilots and RIOs) with a moderate amount of flight experience (900 to 1,500 hours) than among those with a larger amount of experience (1,600 to 2,400 hours) (Lewis & Rimland, 1979).

We have found the asymmetry model to have limited usefulness for personnel assessment. We have found asymmetry in active duty, average ability personnel, but not discharged enlistees, remedial readers, and lower aptitude subjects (Lewis, 1983b). There may be specialization of some hemispheric function, but it has been difficult to show consistent relationships to on-job performance in other areas of our research.

Variability

Perhaps a more appropriate model for our research involves brain variability and its converse, brain stability. Differences in performance are related to the degree to which brain recordings vary. Research in the areas of functional psychiatric disorders (that is, schizophrenia) and effects due to age has made substantial contributions to the development of this model (Callaway, 1975; Callaway & Halliday, 1973; Callaway, Jones, & Donchin, 1970; Shagass, 1972). We have found that "normal" populations also show variability in brain recording. Consistently, we have seen through much of our research that high temporal and spatial variability in the brain is often associated with low performers. High performers generally show less variability and more intra- and inter-subject brain stability than do low performers.

In a study reported earlier (Lewis, Rimland, & Callaway, 1977), we examined a group of recruits ($N = 206$). Half of the recruits were classified as having low aptitude (20–40th centiles on the AFQT), while the other half were classified as having high aptitude (80–99th centiles). Large group differences were found ($t = 2.97$, $df = 204$, $p < .01$) when the trial-to-trial variability measure was used to differentiate between the high and low aptitude groups. The HIGH group showed less variablity ($MN = 7.56 +/-$ 1.38 uVrms) than did the LOW group ($MN = 8.29 +/- 2.03$ uVrms).

In the aviator study, recordings were made from four paired sites: frontal, central, parietal, and occipital. In order to provide general front-to-back comparisons, frontal and central asymmetry values were combined (front), as were parietal and occipital (back). Data were analyzed for the front and back combined sites. In addition we were interested in the dispersion of these measures as they relate to performance rating. The standard deviation statistic, which measures dispersion of a sample, may also provide information about individual and group differences. Such dispersion was assessed in our aviator (Lewis & Rimland, 1979) and sonar operator (Lewis & Rimland, 1980) research. One finding, which has been consistent over several projects, is illustrated in Figure 5.2. This figure shows the standard deviations (*SD*s) plotted for the pilots and RIOs, their performance ratings (high and low), and electrode sites (front and back). Left-handed and ambidextrous subjects were removed from the sample because it was thought that hemisphericity might be mixed in these subjects. The *SD*s for both the high-rated pilots and high-rated RIOs were about equal at the front and back sites, with the *SD*s slightly larger for the back than for the front sites. The *SD*s obtained for the low-rated groups at the front and back sites were greater than those for the corresponding high-rated groups. Further, the *SD*s obtained for low-rated pilots at the front and back sites were greater than those obtained for low-rated RIOs at these sites. As with the high-rated pilot and RIO groups, the *SD*s for the low-rated pilot and RIO groups were greater for the back than the front sites. The front-to-back differences were considerably greater for the low groups.

The back electrode sites included an association area (parietal) and the primary visual reception area (occipital). The front site included both an association area (frontal) and a sensory-motor area (central). The experimental task was passive, requiring only that the subjects observe a blinking light.

An observation in this project dealt with EP habituation—another source of variability (Lewis, 1979). Visual EP habituation was assessed by comparing the EP records from the first 50 flashes with the second 50 flashes. The instructor pilots showed visual EP habituation ($F = 5.98$, $df = 1, 27$, $p < 0.02$), while the student pilots did not. This suggested that perhaps the instructors may have adapted more quickly to the experimental conditions and were less aroused than the students.

Relationships between asymmetry dispersion and performance for the sonar operator trainees were similar to those for the aviators (Figure 5.3). Again, only right-handed subjects (HIGH, $N = 10$, LOW, $n = 10$) were included. The *SD*s, or asymmetry dispersion measures, were similar from the front and back electrode sites for the HIGH group. Greater front-to-back

Figure 5.2
Asymmetry Standard Deviations for the HIGH- and LOW-Rated Pilots and RIOs

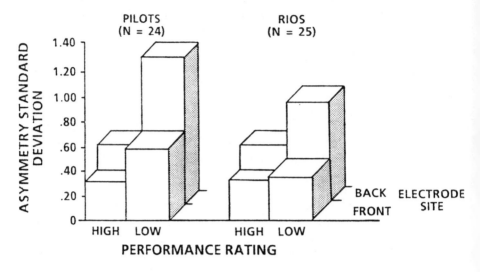

Figure 5.3
Asymmetry Standard Deviations for HIGH- and LOW-Rated Sonar Operator Trainees

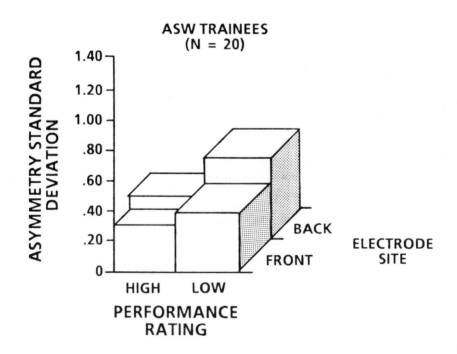

differences were found for the LOW group than for the HIGH group. Finally, there was less dispersion in both front and back regions for the HIGHs compared with the LOWs.

Resource Allocation

The third model, resource allocation (Broadbent, 1958; Coles & Gratton, 1985; Isreal et al., 1980; Wickens, 1980), is also being followed in our research on decision making under varying workload conditions (Trejo, 1986). Each individual has finite resources to devote to a particular activity. The proportion of resources which may be dedicated to an activity or problem may also be assessed by neuroelectric and neuromagnetic procedures. One of the procedures used in neuroelectric assessment of cognitive function involves presenting to the subject an irrelevant probe stimulus, such as a flash of light or clicks to the ears. While performing a task or simulated task, an individual is presented stimuli, which he or she is told to disregard. These stimuli are the irrelevant (to the task) probes used to generate the EP. The probes are most often visual or auditory stimuli. The model suggests that as mental resources are used in performing the task, fewer resources are available for processing the probe stimulus. Consequently, probe-generated EP component amplitude would be expected to decrease and component latency increase as a result of increased resource demands from the primary task. Papanicolaou and Johnstone (1984) have reviewed the use of the irrelevant probe technique and allocation of resources model in cognitive processing.

In a recent report, Trejo (1986) discussed the assumptions, hypotheses, and experimental designs in using neuroelectric, and possibly neuromagnetic, signals as predictors or correlates of decision making by combat system operators. Such decision making would occur under varying workload conditions. He discussed the application and relevance of signal detection theory (Green & Swets, 1966) to the behavioral aspects of the decision-making task. The influence of perceptual sensitivity on signal detection was discussed as was the response bias of individuals performing decision-making tasks. Decision making implies an outcome of the process. Individuals generally assign an expected value of utility to decision alternatives. Trejo discussed the implications of the subjective expected utility theory (Whalen, 1984; Wright, 1984) for the decision-making tasks. An individual's assessment of decision outcome probability may be influenced by personalitylike factors, such as risk-aversion or risk-taking strategies. Trejo states that most decision makers are risk-aversive, which obviously influences the strategy in reaching a decision.

Trejo, Lewis, and Blankenship (1987a) reported on irrelevant probe EP data acquired during the performance of the Air Defense Radar Simulation Task (AIRDEF); (Kelly, Greitzer, & Hershman, 1981; Trejo, 1986). The subject views two concentric circles on a graphics monitor. The outer circle represents the range of a simulated ship-board radar system. Range of the ship's weapon is represented by the inner circle. The subjects are to imagine that they are onboard the ship, represented by a cross at the center of the two concentric circles, incoming hostile missiles appear at any point on the outer circle (radar range) and proceed toward the center (subject's own ship) at one of three speeds (fast, medium, slow). The subjects are required to make a decision as to when to fire antimissile weapons. Obviously, subjects must fire their weapons earlier for a fast incoming missile than for a slow missile. The objectives are to (1) not take "hits" on one's own ship, (2) "kill" all of the incoming missiles at a maximum range, and (3) not fire more than one weapon on the same incoming missile track number. Workload is varied by presenting 18 targets during one condition and 36 during another.

A baseline condition, where no targets were presented, was used to represent resources undiminished by task performance. During the task, the subject observed, but was told to ignore, dim flashes on the graphics monitor. These flashes were pesented aperiodically and provided the visual stimulus to generate the EP epochs for averaging. Forty-five male subject performed the AIRDEF task. Eight channels of EP data were obtained from the frontal, temporal, parietal, and occipital regions of each subject.

As stated earlier, the resource allocation model predicts that as resources are allocated and used to perform a task, fewer resources are available to "process" the irrelevant probe stimulus. One expects, then, that EP amplitudes would decrease from baseline to active-workload conditions. Trejo, Lewis, & Blankenship, (1987a) used two EP measures to assess the effects of workload on decision making during AIRDEF. The first was a traditional signal-averaged EP ($n = 6$ epochs per average) and resulting root mean square (RMS) amplitude (Callaway, 1975; Lewis & Froning, 1981). The metric for this waveform was designated as RMS-a, expressed in microvolts (uV). The second measure was a signal-to-noise ratio, similar to that used by John et al. (1983). The latter waveform was expressed as a ratio of the arithmetic mean to be unbiased standard deviation of corresponding time points for each of the six epochs. The signal-to-noise measure was used also to minimize large random components that may be artifact. The root mean square was obtained for this waveform and was expressed as RMS-s. The latter units were dimensionless because both mean and standard deviation values contained the same units (uV).

Trejo, Lewis, & Blankenship, (1987a) used repeated measures analysis of variance to evaluate the effects of workload (baseline, 18 targets, 36 targets), recording sites, and time windows within each EP. The two workload conditions translated to 4.5 and 9 targets per minute for the 18 and 36 targets, respectively. Recording sites included those from the front (F3, F4), temporal (T3, T4), parietal (P3, P4), and occipital (O1, O2) regions. Because no hemisphere-related differences were found, the mean of the homologous site RMS values was used. Eight time windows were analyzed within the EP waveforms, each about 50 ms wide, extended from about 50 ms through 450 ms. RMS values were computed for each window.

Results showed that workload did decrease the amplitude (RMS-s) of the EP waveform by about 25 percent when compared with baseline ($F = 10.97$, $df = 2, 58$, $p < .001$), as predicted by the resource allocation model. Main effect for site was also highly statistically significant ($F = 22.38$, $df = 3, 87$, $p < .001$), as was the main effect for time window ($F = 10.48$, $df = 7, 203$, $p < .001$). A three-way interaction (workload × site × time window) was significant ($F = 2.44$, $df = 42, 1218$, $p < .001$). This interaction suggested that certain components (time windows) at particular sites were sensitive to workload. Specifically, frontal amplitude decreased by 47 percent in the latency interval 100–150 ms ($F = 54.75$, $df = 1, 5664$, $p < .001$), by 38 percent in the 250–300 ms interval ($F = 20.84$, $df = 1, 5664$, $p < .001$), and 39 percent between 300–350 ms ($F = 23.66$, $df = 1, 5664$, $p < .001$). Amplitudes in the parietal region decreased 41 percent during the 200–250 ms interval ($F = 48.74$, $df = 1, 5664$, $p < .001$). Occipital amplitudes decreased 8 percent in the 100–150 ms interval ($F = 6.58$, $df = 1, 5664$, $p < .025$), and decreased 29 percent between 200–250 ms ($F = 24.04$, $df = 1, 5664$, $p < .001$). Similar effects were noted for the traditional amplitude measure (RMS-a); however, no main effect for workload was found. Frontal amplitudes *decreased* 40 percent during the 100–150 ms interval ($F = 36.54$, $df = 1, 5664$, $p < .001$), 33 percent between 250–300 ms ($F = 17.85$, $df = 1, 5664$, $p < .001$), and 40 percent during the 300–350 ms interval ($F = 30.18$, $df = 1, 5664$, $p < .001$). Parietal amplitude *decreased* 46 percent between 250–250 ms ($F = 53.83$, $df = 1, 5664$, $p < .001$); however, amplitude in the occipital region *increased* 13 percent between 100–150 ms ($F = 10.63$, $df = 1, 5664$, $p < .001$) and *decreased* 29 percent between 200–250 ms ($F = 51.32$, $df = 1, 5664$, $p < .001$).

Relationships between these data and on-job performance have also been found (Trejo, Lewis, & Blankenship, 1987b). On-job performance data dealt with military and job knowledge and performance, reliability, and motivation. Two groups (HIGH and LOW) were formed based on these data. The mean age for the HIGH group ($n = 16$), was 21 +/− 2 years, while that for the LOW group ($n = 10$) was 20 +/− 1 years.

EP amplitude measures (RMS-a) showed a relationship between on-job performance and workload effects. The two groups had statistically significant amplitude differences (workload minus baseline) at two windows centered at 175 ms and 325 ms. There was about a 2 uVrms decrease from the workload and baseline conditions for the HIGH group and little or no decrease for the LOW group. One interpretation may be that individuals in the HIGH group may be better able to shift resources from the probe to the task engagement than those in the LOW group.

NEW TECHNIQUES FOR MEASURING BRAIN ACTIVITY

For the past several years, NPRDC has been developing neuromagnetic (evoked fields, EF) recording capability to assess brain processing and "index" on-job performance. These recordings have several advantages over traditional neuroelectric procedures. (1) They represent an absolute measure, compared with the EEG/EP, which is a relative measure between an active site and a relatively "indifferent" reference electrode. Activity at the reference site often complicates interpretation of the EPA data and makes precise location of brain activity difficult. (2) With EF recordings, there is little or no effect on, or "smearing" of, the neuromagnetic field due to skull capacitance or skull-scalp tissue interface. EF activity, therefore, has higher spatial resolution than does EP activity. (3) The EF may also provide information in addition to what may be obtained using the EP. Because of these advantages, we may be able to increase our capability to assess individual differences, and, therefore, predict on-job performance (Lewis, 1983a; Lewis & Blackburn, 1984). Improving technology and methodology in order to obtain single epochs (unaveraged records) would provide more accurate assessment of short-term brain processing. Two years ago we reported on EF single epoch recordings as well as test-retest EF reliability, the first such reports to appear in the literature (Lewis et al., 1985). Recent research at the Center has suggested that neuromagnetic recordings may predict on-job performance better than neuroelectric recordings.

EP and EF data were recently obtained in a military operational environment for the first time and EP and EF relationships were found with on-job performance (Lewis et al., 1986). These data were obtained from 26 Marine Corps personnel. On-job performance criteria data were obtained by supervisor ratings and dealt with military and job knowledge and performance, reliability, and motivation. The supervisor rated each subject as "high," "satisfactory," or "low" for each of the above criteria. Two groups (HIGH and LOW) were determined from the ratings. The criterion for assignment to the HIGH group was "high" ratings in all categories. One or more ratings of less than "high" resulted in assignment to the LOW group. The entire sample of subjects should be considered fairly homogeneous because they all were similar in age, had been in the military for four years, were all males, and were highly selected for security positions.

Correlation between the HIGH:LOW job performance rating given in 1985 and their rank in 1987 based on Marine Corps records was .079 ($p < .0002$. To the extent that as rank indicates on-job performance quality, the HIGH:LOW job rating may be considered a reasonable index of performance. The mean age for the HIGH group ($n = 16$) was 21 $+/- 2$ years, while that for the LOW group ($n = 10$) was 20 $+/- 1$ years. Data were also obtained from two standardized tests, the Cognitive Laterality Battery (CLB) (Gordon, 1983) and the Test of Attention and Interpersonal Style (TAIS) (Nideffer, 1977). The CLB is a series of tests that assess cognitive functions such as verbal/sequential and visuospatial processing. It has been used with such diverse occupational groups as combat pilot trainees, bank employees, and computer programmers. The TAIS is a self-report inventory that assesses the respondent's ability to control attention and interpersonal factors. Such factors have been suggested as important in emergency situations, competitive athletics, and business. This test has been used as a personnel selection battery for occupations, incuding police officers.

Each subject viewed (binocular, central fixation) a black-and-white checkerboard pattern subtending 5 degrees visual angle (VA) at a luminance of about 34 cd/sqm. Each check subtended .04 degrees VA. The stimulus was flashed on for 10 msec. Intertrial interval varied between 500 and 1500 msec. Background luminance was about 3 cd/sqm. EP data were recorded using a commercially available electrode helmet (Electro-Cap International[2]), amplified (20,000 gain), and bandpassed (0.1–100 Hz; Grass amplifiers model 12A5). Ten channels of data were obtained; however, data from only two sites (visual reception/occipital area 01, 02) will be discussed here. EF recordings were obtained using a DC SQUID Biomagnetic Detection System (B.T.I., Inc. model 600B, second derivative gradiometer). The single channel EF signal (1000 gain on the SQUID control unit) was bandpassed (0.1–40 Hz Krohn-Hite, mode 3343) and further amplified (50 gain, Grass P511J) prior to digital conversion. Sampling rate for the EP and EF recording was 256 Hz. Post-stimulus record lengths were one-half second. EPS were averaged over 7 epochs, while EFs were averaged over 19 epochs.

All EP and EF data were acquired and stored as single epochs on a field-portable computer system (MASSCOMP, model MCS-5500). The unit of measure for the EPs was the microvolt (uV), while that for the EF was the femtotesla (fT) (10^{-15} Tesla). Sample EP and EF data recorded over the left (O1) and right occipital (O2) areas appear in Figure 5.4. More precisely, the EF data were recorded 1 cm lateral to the O1 and O2 EP sites. However, for convenience, the EF site locations will be referred to as O1 and O2. Note the similiarity in the EP data recorded over the two separate areas (O1 versus O2) and the polarity reversal in the EF data recorded from the same general areas (O1 versus O2). For both the EP and EF records, root mean square (rms) amplitudes were obtained from each single epoch and the averaged data.

Neither the TAIS nor the CLB scores were able to distinguish the HIGH group from the LOW group (t-test). The CLB did show that both groups of security personnel performed better on tests of visuospatial function than on verbal/sequential tests (nonparametric sign tests, $p < .02$). This finding suggests that the subjects were a homogeneous sample.

Mean, standard deviation, coefficient of variation ($CV = SD/MN$), and t-test data for EP and EF recordings, sites, and performance groups appear in Table 5.3. Three LOW group and two HIGH group subjects lacked EP data, reducing the group sizes to $n = 7$ and $n = 14$ for the LOW and HIGH groups, respectively. The LOW group had lower amplitudes than did the HIGH group for EP and EF recordings over both sites (O1 and O2). Evoked potential CVs were about the same for both LOW and HIGH groups; however, they were much greater for the LOW than for the HIGH group at

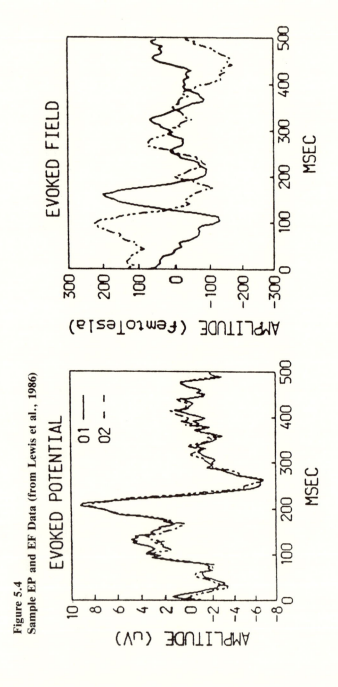

Figure 5.4
Sample EP and EF Data (from Lewis et al., 1986)

Table 5.3
Descriptive and Inferential Statistics for Evoked Potentials and Evoked Fields, Sites, and Performance Groups

	Evoked Potentials (μVrms)				Evoked Fields (fTrms)			
	Site 01		Site 02		Site 01		Site 02	
	Low (N=7)	High (N=14)	Low (N=7)	High (N=14)	Low (N=10)	High (N=16)	Low (N=10)	High (N=16)
MN	5.00	6.99	4.83	6.73	217	272	173	331
SD	1.70	2.20	1.71	2.47	145	105	115	119
CV	.34	.31	.35	.37	.67	.39	.66	.36
t		2.08		1.81		1.12		3.33
df		19		19		24		24
p		.05		.08		.27		.003

both EF recording sites. Group differences were found for the EP data at site O1 ($p < .05$) and for the EF data at site O2 ($p < .003$). Even though the EF *SD*s were about the same for both groups at site O2, the mean value for the HIGH group was nearly two times that for the LOW group (Table 5.3, Figure 5.5). Largest group differences were seen at the EF site O2, which is reflected in the large *t*-test value and *p* value in Table 5.3.

EP data recorded over the left hemisphere occipital (that is, vision reception) area were able to statistically differentiate the HIGH from the LOW groups ($t = 2.08$, $df = 19$, $p < .05$). Those EP data recorded over the right hemisphere did not show statistically significant group differences. EF data were able to show performance group differences to a much greater degree than did the EP data ($t = 3.33$, $df = 24$, $p < .003$). EF temporal and spatial variability was found to be greater for the LOW group than for the HIGH group, a finding that supports the results from our earlier EP research. Our data suggest that inter-individual and inter-group differences may be more pronounced with EF recordings than with EP. Improved localization of the EF recording over the EP recording may, in part, account for increased sensitivity to individual and group differences. Both the EF and EP finding showed group differences that were not seen by either the CLB or TAIS tests. Neither the CLB nor the TAIS test scores correlated with job performance.

In conclusion, the Navy Personnel Research and Development Center has been exploring the use of neuroscience technologies to improve personnel assessment. Such assessment includes improving the prediction of on-job peformance, fitness for duty, selection, and classification. Other research areas include information processing during decision making. Various levels of workload are being used during the performance of realistic simulations to assess decision making. Research is continuing in the area of personnel reliability. Recent operational-recorded EF findings demonstrate

Figure 5.5
Group Mean Values for EP and EF Data at Sites 01 and 02 (HIGH Group Drawn in Solid Lines, LOW Group in Dashed Lines)

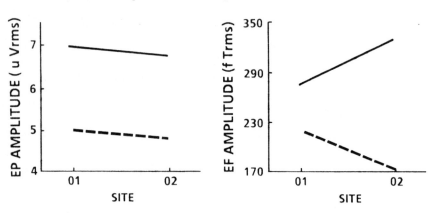

that neuromagnetic recording can be obtained outside of a highly controlled laboratory environment and can be related to on-job performance (Lewis et al., 1987).

NOTES

1. It should be pointed out that Vetterli and Furedy incorrectly referenced the Blinkhorn and Hendrickson study as showing EP/IQ correlations between 0.7 and 0.8. Correct correlation (0.54) is cited above. It was the Hendrickson and Hendrickson (1980) study that provided the correlations ranging between 0.7 and 0.8.

2. Identification of equipment is for documentation only and does not imply endorsement.

REFERENCES

Beaumont, J. G., Young A. W., and McManus, I. C. (1984). Hemisphericity: A critical review. *Cognitive Neuropsychology, 1,* 191–212.

Bennett, W. F. (1968). Human perception: Network theory approach. *Nature, 220,* 1147–1148.

Berger, H. (1933). Uber das Electroenkephalogram des Menschen. *Archiv fur Psychiatrie und Nervenkrankheiten, 98,* 231–254.

Berkhout, J., and Walter, D. O. (1968). Temporal stability and individual differences in the human EEG: An analysis of variance of spectral values. *IEEE Transactions on Bio-Medical Engineering, BME-15,* 165–168.

Blinkhorn, S. F., and Hendrickson, D. E. (1982). Averaged evoked responses and psychometric intelligence. *Nature, 295,* 596–597.

Boring, E. G. (1950). *A history of experimental psychology.* New York: Appleton-Century-Crofts.

Brazier, M. A. B. (1962). The analysis of brain waves. *Scientific American, 206,* 142–153.

Broadbent, D. (1958), *Perception and communication.* Oxford: Pergamon Press.

Buchsbaum, M. and Fedio, P. (1969). Visual information and evoked responses from the left and right hemispheres. *Electroencephalography and Clinical Neurophysiology, 26,* 266–272.

Buchsbaum, M. S., and Coppola, R. (1977). Signal-to-noise ratio and response variability in affective disorders and schizophrenia. In H. Begleiter (Ed.), *Evoked brain potentials and behavior* (pp. 447–465). New York: Plenum Press.

Buchsbaum, M. S., Henkin, R. J., and Christiansen, R. L. (1974). Age and sex differences in averaged evoked responses in a normal population, with observations on patients with gonadal dysgenesis. *Electroencephalography and Clinical Neurophysiology, 37,* 137–144.

Buchsbaum, M., and Pfefferbaum, A. (1971). Individual differences in stimulus intensity response. *Psychophysiology, 8,* 600–611.

Buss, A. R., and Poley, W. (1976). *Individual differences: Traits and factors.* New York: Gardner Press.

Callaway, E. (1975). *Brain electrical potentials and individual psychological differences.* New York: Grune & Stratton.

Callaway, E., and Halliday, R. A. (1973). Evoked potential variability: Effects of age, amplitude and methods of measurement. *Electroencephalography and Clinical Neurophysiology, 34,* 125–133.

Callaway, E., Jones, R. T., and Donchin, E. (1970). Auditory evoked potential variability in schizophrenia. *Electroencephalography and Clinical Neurophysiology, 29,* 421–428.

Chalk, F. C. R., and Ertl, J. (1965). Evoked potentials and intelligence. *Life Sciences, 4,* 1319–1322.

Cohen, R. (1972). The influence of task-irrelevant stimulus-variations on the reliability of auditory evoked responses in schizophrenia. In A. Fessard and G. Lelord (Eds.), *Human neurophysiology, psychology, and psychiatry.* INSERM Colloquium on "Averaged Evoked Responses and their Conditioning in Normal Subjects and Psychiatric Patients." (Tours, France, 22–23 Sept. 1972).

Coles, M. G. H., and Gratton, G. (1985). Psychophysiology and contemporary models of human information processing. In D. Papakostopoulos, S. Butler, and I. Martin (Eds.), *Clinical and experimental neuropsychophysiology.* London, Eng.: Croom Helm.

Dahlstrom, W. G. (1985). The development of psychological testing. In G. A. Kimble and K. Schlesinger (Eds.), *Topics in the history of psychology: Vol. 2* (pp. 63–113). Hillsdale, NJ: Erlbaum Associates.

Davis, F. B. (1971). *The measurement of mental capability through evoked potential recordings.* Greenwich, CT: Educational Records Bureau.

Davis, H., and Davis, P. A. (1936). Action potentials of the brain in normal persons and in normal states of cerebral activity. *Archives of Neurology and Psychiatry, 36,* 1214–1224.

Dillon, R. F., and Wisher, R. A. (1981). The predictive validity of eye movement indices for technical school qualifying test performance. *Applied Psychological Measurement, 5*(1), 43–49.

Dimond, S. J., and Beaumont, J. G. (Eds.). (1974). *Hemisphere function in the human brain.* New York: John Wiley.

Doktor, R., and Bloom, D. M. (1977). Selective lateralization of cognitive style related to occupation as determined by EEG alpha asymmetry. *Psychophysiology, 14*, 385–387.

Dustman, R. E., and Beck, E. C. (1963). Long-term stability of visually evoked potentials in man. *Science, 142*, 1480–1481.

_____ (1969). The effects of maturation and aging on the wave form of visually evoked potentials. *Electroencephalography and Clinical Neurophysiology, 26*, 2–11.

_____ (1972). Relationship of intelligence to visually evoked responses. *Electroencephalography and Clinical Neurophysiology, 33*, 254.

Ellingson, R. J. (1970). Variabiity of visual evoked responses in the human newborns. *Electroencephalography and Clinical Neurophysiology, 29*, 10–19.

Ellingson, R. J., Danahy, T., Nelson, B., and Lathrop, G. H. (1974). Variability of auditory evoked potentials in human newborns. *Electroencephalography and Clinical Neurophysiology, 36*, 155–162.

Ellingson, R. J., Lathrop, G. H., Danahy, T., and Nelson, B. (1973). Variability of visual evoked potentials in human infants and adults. *Electroencephalography and Clinical Neurophysiology, 34*, 113–124.

Ertl, J. P. (1968). *Evoked potentials, neural efficiency and I.Q.* Paper presented at the International Symposium for Biocybernetics, Washington, D.C.

_____ (1969). *Neural efficiency and human intelligence* (Final Report) (Project No. 9–0105). Washington, DC: Department of Health, Education, & Welfare.

_____ (1971). Fourier analysis of evoked potentials and human intelligence. *Nature, 230*, 525–526.

_____ (1973). I.Q., evoked responses and Fourier analysis. *Nature, 241*, 209–210.

Ertl, J. P., and Schafer, E. (1969). Brain response correlates of psychometric intelligence. *Nature, 223*, 421–422.

Everhart, J. D., China, C. L., and Auger, R. A. (1974). Measures of EEG and verbal intelligence: An inverse relationship. *Physiological Psychology, 2*, 374–378.

Eysenck, H. J., and Barrett, P. (1985). Psychophysiology and the measurement of intelligence. In C. R. Reynolds and V. L. Willson (Eds.), *Methodological and statistical advances in the study of individual differences* (pp. 1–49). New York: Plenum Press.

Fein, G., Galin, D., Johnstone, J., Yingling, C. D., Marcus, M., and Kiersch, M. E. (1983). EEG power spectra in normal and dyslexic children. I. Reliability during passive conditions. *Electroencephalography and Clinical Neurophysiology, 55*, 399–405.

Galbraith, G. C., Gliddon, J. B., and Busk, J. (1970). Visual evoked responses in mentally retarded and nonretarded subjects. *American Journal of Mental Deficiency, 75*, 341–348.

Galin, D., and Ellis, R. R. (1975). Asymmetry in evoked potentials as an index of lateralized cognitive processes: Relation to EEG alpha asymmetry. *Neuropsychologia, 13*, 45–50.

Galin, D., and Ornstein, R. (1974). Individual differences in cognitive style: I. Reflective eye movements. *Neuropsychologia, 12*, 367–376.

Ghiselli, E. E. (1966). *The validity of occupational aptitude tests.* New York: John Wiley.

_____ (1973). The validity of aptitude tests in personnel selection. *Personnel Psychology, 26*, 461–477.

Glaser, E. M., and Ruchkin, D. S. (1976). *Principles of neurobiological signal analysis.* New York: Academic Press.

Gordon, H. W. (1983). *The Cognitive Laterality Battery.* Pittsburgh: University of Pittsburgh, Western Psychiatric Institute and Clinic, Media Productions Services.

Green, D. M., and Swets, J. A. (1966). *Signal detection theory and psychophysics.* New York: John Wiley.

Hadley, J. M. (1940). Some relationships between electrical signs of central and peripheral activity: I. During rest. *Journal of Experimental Psychology, 27*, 640–656.

_____ (1941). Some relationships between electrical signs of central and peripheral activity: II. During mental work. *Journal of Experimental Psychology, 28*, 53–62.

Helvey, T. C. (1975). Learning disability measurement with the synchrocephalograph. *Journals of Experimental Education, 44*, 18–25.

Hendrickson, D. E., and Hendrickson, A. E. (1980). The biological basis of individual differences in intelligence. *Personality and Individual Differences, 1*, 3–33.

Henry, C. E. (1941a). Electroencephalographic individual differences and their constancy: I. During sleep. *Journal of Experimental Psychology, 29*, 117–132.

_____ (1941b). Electroencephalographic individual differences and their constancy: II. During waking. *Journal of Experimental Psychology, 29*, 236–247.

Hoagland, H. Cameron, D. E., Rubin, M. A., and Tegelberg, J. J. (1938). Emotion in man as tested by the delta index of the electroencephalogram: II. Simultaneous records from cortex and from a region near the hypothalamus. *Journal of General Psychology, 29*, 247–261.

Hunt, E., and Pellegrino, J. (1985). Using interactive computing to expand intelligence testing: A critique and prospectus. *Intelligence, 9*, 207–236.

Hunter, J. E., and Hunter, R. F. (1984). Validity and utility of alternative predictors of job performance. *Psychological Bulletin, 96*, 72–98.

Isreal, J. B., Chesney, G. L., Wickens, C. D., and Donchin, E. (1980). P300 and tracking difficulty: Evidence for multiple resources in dual-task performance. *Psychophysiology, 17*, 259–273.

Jasper, H. (1958). The ten-twenty electrode system of the International Federation. *Electroencephalography and Clinical Neurophysiology, 10*, 371–375.

Jensen, A. (1985). Methodological and statistical techniques for the chronometric study of mental abilities. In C. R. Reynolds and V. L. Willson (Eds.), *Methodological and statistical advances in the study of individual differences.* New York: Plenum Press.

John, E. R., Prichep, L., Ahn, H., Easton, P., Fridman, J., and Kaye, H. (1983). Neurometric evaluation of cognitive dysfunctions and neurological disorders in children. *Progress in Neurobiology, 21*, 239–290.

Kelly, R. T., Greitzer, F. L., and Hershman, R. L. (1981). *Air defense: A computer game for research in human performance* (NPRDC Report No. 81–15). San Diego: Navy Personnel Research and Development Center.

Kinsbourne, M. (1972). Eye and head turning indicates cerebral lateralization. *Science, 176*, 539–541.

_____ (Ed.). (1978). *Asymmetrical function of the brain*. New York: Cambridge University Press.

Knights, R. M., and Bakker, D. J. (1976). *The neuropsychology of learning disorders: Theoretical approaches*. Baltimore: University Park Press.

Knott, J. R. (1938). Brain potentials during silent and oral reading. *Journal of General Psychology, 18*, 57-62.

Kooi, K. A., and Bagchi, B. K. (1964). Visual evoked responses in man: Normative data. *Annals, New York Academy of Sciences, 112*, 254-269.

Kreezer, G. (1937). The dependence of the electro-encephalogram upon intelligence level. *Psychological Bulletin, 34*, 769-770.

_____ The electro-encephalogram and its use in psychology. *American Journal of Psychology, 51*, 737-759.

_____ Intelligence level and occipital alpha rhythm in the Mongolian type of mental deficiency. *American Journal of Psychology, 52*, 503-532.

Kreezer, G., and Smith, F. W. (1936). Electrical potentials of the brain in certain types of mental deficiency. *Archives of Neurology and Psychiatry, 36*, 1206-1213.

_____ (1937). Brain potentials in the heredity type of mental deficiency. *Psychological Bulletin, 34*, 535-536.

Larson, G. E., and Saccuzzo, D. P. (1986). Jensen's reaction-time experiments: Another look. *Intelligence, 10*, 231-238.

Lelord G., Laffont, F., and Jusseaume, P. (1976). Conditioning of evoked potentials in children of differing intelligence. *Psychophysiology 13*, 81-85.

Lennox, W. G., Gibbs, E. L., and Gibbs, F. A. (1945). The brain-wave pattern, an hereditary trait. *Journal of Heredity, 36*, 233-243.

Lewis, G. W. (1979). Visual event related potentials of pilots and navigators. In D. Lehmann and E. Callaway (Eds.), *Human evoked potentials. Applications and problems* (p. 462). New York: Plenum Press.

_____ (1983a). Event related brain electrical and magnetic activity: Toward predicting on-job performance. *International Journal of Neuroscience, 18*, 159-182.

_____ (1983b). *Bioelectric predictors of personnel performance: A Review of relevant research at the Navy Personnel Research and Development Center*. (NPRDC Report No. 84-3). San Diego: Navy Personnel Research and Development Center.

_____ (1984). Temporal stability of multichannel, multimodal ERP recordings. *International Journal of Neuroscience, 25*, 131-144.

Lewis, G. W., and Blackburn, M. R. (1984). *Biomagnetism: Possible new predictor of personnel performance* (NPRDC Report No. 84-43). San Diego: Navy Personnel Research and Development Center.

Lewis, G. W., Blackburn, M. R., Naitoh, P., and Metcalfe, M. (1985). Few-trial evoked field stability using the DC SQUID. In H. Weinberg, G. Stroink, and T. Katila (Eds.), *Biomagnetism: Applications and theory* (pp. 343-347). New York: Pergamon Press.

Lewis, G. W., and Froning, J. N. (1981). Sensory interaction, brain activity, and reading ability in young adults. *International Journal of Neuroscience, 15*, 129-140.

Lewis, G. W., and Rimland, B. (1979). *Hemispheric asymmetry as related to pilot and radar intercept officer performance* (NPRDC Report No. 79-13). San Diego: Navy Personnel Research and Development Center.

_____ (1980). *Psychobiological measures as predictors of sonar operator performance* (NPRDC Report No. 80-26). San Diego: Navy Personnel Research and Development Center.

Lewis, G. W., Rimland, B., and Callaway, E. (1976). *Psychobiological predictors of success in a Navy remedial reading program.* (NPRDC Report No. 77-13). San Diego: Navy Personnel Research and Development Center.

_____ (1977). *Psychobiological correlates of aptitude among Navy recruits* (NPRDC Note 77-7). San Diego: Navy Personnel Research and Development Center.

Lewis, G. W., Trejo, L. J., Blackburn, M. R., and Blankenship, M. H. (1986). Neuroelectric and neuromagnetic recordings: Possible new predictors of on-job performance. In G. E. Lee (Ed.), *Proceedings: Psychology in the Department of Defense. Tenth Symposium* (pp. 606-610). Colorado Springs: U.S. Air Force Academy (USAFA TR 86-1).

Lewis, G. W., Trejo, L. J., Nunez, P. L., Weinberg, H., and Naitoh, P. (1987, August). *Evoked neuromagnetic fields: Implications for indexing performance.* Presented at the 6th International Conference on Biomagnetism, Tokyo, Japan.

Linn, R. (1982). Ability testing: Individual differences, prediction, and differential prediction. In A. K. Wigdor and W. R. Garner (Eds.), *Ability testing: Uses, consequences, and controversies. Part II: Documentation Section* (pp. 335-388). Washington, DC: National Academy Press.

Lykken, D. T. (1973). The "neural efficiency analyzer" scandal. *Contemporary Psychology, 18,* 462-463.

Matousek, M., Arviddson, A., and Friberg, S. (1979). Serial quantitative electroencephalography. *Electroencephalography and Clinical Neurophysiology, 47,* 614-622.

Mintzberg, H. (1976). Planning on the left side and managing on the right. *Harvard Business Review, 54,* 49-58.

Mitchell, J. V., Jr. (1985). *The ninth mental measurements yearbook.* Lincoln, NE: The Buros Institute of Mental Measurement.

Nideffer, R. M. (1977). *Test of attentional and interpersonal style.* San Diego: Enhanced Performance Associates.

Ornstein, R. (1977). *The psychology of consciousness* (2nd ed.). New York: Harcourt Brace Jovanovich.

_____ (1978). The split and the whole brain. *Human Nature, 1,* 76-83.

Papanicolaou, A. C., and Johnstone, J. (1984). Probe evoked potentials: Theory, method, and applications. *International Journal of Neuroscience, 24,* 107-131.

Rhodes, L. E., Dustman, R. E., and Beck, E. C. (1969). The visual evoked response: A comparison of bright and dull children. *Electroencephalography and Clinical Neurophysiology, 27,* 364-372.

Rubin, M. A. (1938). A variability study of the normal and schizophrenic occipital alpha rythm. *The Journal of Psychology, 6,* 325-334.

Sacuzzo, D. P., Larson, G. W., and Rimland, B. (1986). *Speed of information processing and individual differences in intelligence* (Report No. 86-25). San Diego: Navy Personnel Research and Development Center.

Shagass, C. (1972). *Evoked brain potentials in psychiatry.* New York: Plenum Press.

Shipley, T. (1980). *Sensory integration in children: Evoked potentials and intersensory functions in pediatrics and psychology.* Springfield, IL: C. C. Thomas.

Shucard, D. W., and Callaway, E. (1973). Relationship between human intelligence and frequency analysis of cortical evoked responses. *Perceptual and Motor Skills, 36,* 147–151.

Shucard, D. W., and Horn, J. L. (1972). Evoked cortical potentials and measurement of human abilities. *Journal of Comparative and Physiological Psychology, 78,* 59–68.

Stassen, H. H. (1980). Computerized recognition of persons by EEG spectral patterns. *Electroencephalography and Clinical Neurophysiology, 49,* 190–194.

Travis, L. E. (1937). Brain potentials and the temporal course of consciousness. *Journal of Experimental Psychology, 21,* 302–309.

Travis, L. E., and Egan, J. P. (1938). Increase in frequency of the alpha rhythm by verbal stimulation. *Journal of Experimental Psychology, 23,* 384–393.

Travis, L. E., and Gottlober, A. (1936). Do brain waves have individuality? *Science, 84,* 532–533.

_____ (1937). How consistent are an individual's brain potentials from day to day? *Science, 85,* 223–224.

Trejo, L. J. (1986). *Brain activity during tactical decision-making: I. Hypotheses and experimental design* (Tech. Note No. 71–86–6). San Diego: Navy Personnel Research and Development Center.

Trejo, L. J., Lewis, G. W., and Blankenship, M. H. (1987a). *Brain activity during tactical decision-making: II. Probe-evoked potentials and workload* (NPRDC Tech. Note No. 87–xx, in press). San Diego: Navy Personnel Research and Development Center.

_____ (1987b). *Brain activity during tactical decison-making: III. Relationships between physiology, simulation performance, and job performance* (NPRDC Tech. Note. No. 87–xx, in preparation). San Diego: Navy Personnel Research and Development Center.

Uttal, W. R., and Cook, L. (1964). Systematics of the evoked somatosensory cortical potential: A psychophysical-electrophysiological comparison. *Annals of the New York Academy of Sciences, 112,* 60–80.

Van Dis, H., Corner, M., Dapper, R., Hanewald, G., and Kok, H. (1979). Individual differences in the human electroencephalogram during quiet wakefulness. *Electroencephalography and Clinical Neurophysiology, 47,* 87–94.

Vetterli, C. F., and Furedy, J. J. (1985). Evoked potential correlates of intelligence: Some problems with Hendrickson's string measure of evoked potential complexity and error theory of intelligence. *International Journal of Psychophysiology, 3,* 1–3.

Vineberg, R., and Joyner, J. N. (1982). *Prediction of job performance: Review of military studies* (Report No. 82–37). San Diego: Navy Personnel Research and Development Center.

Vogel, W., and Broverman, D. M. (1964). Relationship between EEG and test intelligence: A critical review. *Psychological Bulletin, 62*(2), 132–144.

Wasman, M., and Gluck, H. (1975). Recovery functions of somatosensory evoked responses in slow learners. *Psychophysiology, 12,* 371–376.

Weinberg, H. (1969). Correlation of frequency spectra of averaged visual evoked potentials with verbal intelligence. *Nature, 224,* 813-815.

Werre, P. F., and Smith, C. J. (1964). Variability of responses evoked by flashes in man. *Electroencephalography and Clinical Neurophysiology, 17,* 644-652.

Whalen, T. (1984). Decision making under uncertainty with various assumptions about available information. *IEEE Transactions on Systems, Man and Cybernetics, SMC-14,* 888-900.

Wicke, J. D., Donchin, E., and Lindsley, D. B. (1964). Visual evoked potentials as a function of flash luminance and duration. *Science, 146,* 83-85.

Wickens, C. (1980). The structure of attentional resources. In R. Nickerson and R. Pew (Eds.), *Attention and performance VIII.* New Jersey: Erlbaum.

Williams, A. C. (1939). Some psychological correlates of the electroencephalogram. *Archives of Psychology, 240,* 48.

Wright, G. (1984). *Behavioral decision theory.* Beverly Hills: Sage Publications.

6 New Concepts in Large-Scale Achievement Testing: Implications for Construct and Incremental Validity

Karen Janice Mitchell

INTRODUCTION

The information presented in this chapter differs from that of previous sections. The focus here is on innovative, large-scale achievement testing programs. This review looks at efforts to expand the testing domain by federal, state, and commercial education agencies. Measurement of such abilities as critical thinking skills by the Connecticut, Illinois, California, and several other state education departments, problem-solving abilities by the General Management Admission Test and the National Board of Medical Examiners, writing ability by over 30 state agencies and by the Medical College Admission Test of English as a Foreign Language, and logical and analytical reasoning in Florida and New Jersey and by the Graduate Record Examination program has been introduced. New directions for the National Assessment of Educational Progress will also be reviewed.

Discussion focuses on the constant validity of these recently introduced tests and on evidence to date for the predictive validity of such measures to academic performance. Where data are available, conclusions are drawn about the incremental validity of these and similar expansions of the predictor domain in large-scale testing.

TESTING OBJECTIVES

The objectives of large-scale achievement testing programs are many and varied. They include the provision of: (1) diagnostic/prescriptive information about students' strengths and weaknesses; (2) promotion/retention information; (3) accountability information about students, teachers, programs, and systems; (4) selection and classification data; and (5) data for the development of curricula and educational policy. Because the information needs of many federal, state, and commercial programs are extensive, the assessment instruments and testing plans must often, of necessity, compromise the specific requirements of individual objectives in order to be generally valuable. For this reason, the constructs discussed here are more global than those discussed in other chapters; they generally ignore elementary information processing components and include aggregated abilities or skill clusters.

CRITICAL THINKING AND PROBLEM-SOLVING SKILLS

In 1982, the Education Commission of the States reported that American students do not perform well on tasks that require them to apply their knowledge and use higher order thinking skills; the Commission stated that students do not analyze, synthesize, evaluate, or manipulate information well (Education Commission of the States, 1982). The report of the 1983 National Science Board Commission drew similar conclusions (National Science Board Commission, 1983). Many state agencies have since mounted programs to teach and assess the complex cognitive skills needed for critical thinking and problem solving. Research and testing programs have been instituted through the Connecticut Assessment of Educational Progress (Ennis, 1985), the California Assessment Program (Kneedler, 1985), the Illinois Critical Thinking Skills Project (Ennis, 1985), and the Montana, North Carolina, Michigan, and Pennsylvania state education departments (Education Commission of the States, 1984). Research and testing efforts for these programs concentrate on the abilities to gather information, evaluate and manipulate data, and render judgments or actions. Tested skills include the following: (1) identifying central issues or problems, (2) defining and clarifying essential elements or terms of an argument, (3) identifying the structure of a stated problem, (4) distinguishing fact, opinion, and reasoned judgment, (5) judging the credibility of a source, (6) testing conclusions or hypotheses, (7) recognizing the adequacy of data, and (8) presenting a position clearly and concisely in oral or written form.

In each of the state programs, skills are measured in the context of history/social science, reading, literature, science, writing, and/or mathematics curricula. In two of the programs, distinctions are drawn between cognitive and metacognitive abilities. Testing occurs in the state programs somewhere between the fourth grade and second undergraduate year.

Three different approaches to the assessment of critical thinking and problem-solving skills have been adopted by these programs. They employ multiple-choice questions, vocabulary items, and student writing samples. In the case of multiple-choice questions, story problems, tabular data or pictures with background information are typically provided and critical thinking or problem-solving facility is demonstrated by the selection of options. A history/social science item from the California assessment of critical thinking skills reads, for example:

Suppose you know that 80 percent of the people of West Germany are in favor of having cruise missiles in West Germany and that 99 percent of the people at a meeting in Munich are West Germans. Suppose you also know that Hans Planck is at this meeting.

Which of the following would be TRUE?

A. Hans Planck is in favor of having cruise missiles in West Germany.
B. Hans Planck is probably against having cruise missiles in West Germany.
C. Hans Planck might or might not be in favor of having cruise missiles in West Germany.
D. Hans Planck is definitely against having cruise missiles in West Germany.

This item tests the ability to make distinctions between verifiable and unverifiable, relevant and nonrelevant, and essential and incidental information (Kneedler, 1985).

With vocabulary items, terms associated with particular thinking skills, for example, "evaluate," "hypothesize," "analyze," "imply," and "infer," are typically embedded in text and students are asked to identify appropriate meanings for the terms. In late 1986, the California program introduced a writing sample test of critical thinking skills. In this program centrally derived prompts are administered and local scoring is used to evaluate critical thinking skills demonstrated by writing samples.

Though not described as critical thinking or problem-solving abilities, similar characteristics are tested by the Skills Analysis tests of the Medical College Admission Test (MCAT; Association of American Medical Colleges, 1984). The Skills Analysis: Reading and Skills Analysis: Quantitative items assess the abilities to comprehend, evaluate, and use information presented in narrative and quantitative form. The following is a sample item from the Skills Analysis: Quantitative test.

A random sample was taken of 600 female adults. Each individual was asked to recall how many times she had visited a physician in the last year. The following table gives the number of women according to age group who reported 0, 1, 2, or 3 physician visits. No one reported more than 3 visits.

Females

Number of Visits	Age			
	21-40	*41-64*	*65+*	**Total**
0	50	50	25	125
1	0	100	25	125
2	50	50	75	175
3	100	0	75	175
Total	200	200	200	600

The following statement is related to the information presented above:

Females 41-64 years of age averaged 1 reported physician visit per person per year.

Based on the information given, select

A. if the statement is supported by the information given.

B. if the statement is contradicted by the information given.

C. if the statement is neither supported nor contradicted by the information given.

This item tests the ability to recognize relationships between available information and possible conclusions (Association of American Medical Colleges, 1984).

The General Management Admission Test (GMAT) includes a series of verbal items that call for problem solving and situational analysis. Passages describing problem situations in business contexts with alternate courses of action are followed by statements to be categorized as (1) major objectives, (2) major factors, (3) minor factors, (4) major assumptions or (5) unimportant issues (Comer, 1986). Each of the Law School Admission Test, Advanced Placement Test, and Scholastic Aptitude and American College Tests have items that call for multiple abilities including critical thinking skills.

The National Board of Medical Examiners (NBME) administers written clinical simulation exams to assess problem-solving ability for patient management (Hixon & Loadman, 1986). These examinations are used for physician credentialing. They require examinees to select appropriate patient data from a list of alternatives, to synthesize obtained information, and provide a clinical judgment. The NBME is currently examining the psychometric characteristics of computer-based clinical simulation tests.

The content specifications for instructional development and item construction for most of these programs are quite impressive. The California Critical Thinking Skills Continuum (Kneedler, 1985) which details the acquisition of and context for observing thinking skills is a good example of the care involved in the development of program objectives and materials. The Student Performance Standards of Excellence for Florida Schools is also exemplary (Florida Department of Education, 1984). Internal consistency estimates and scorer agreement levels for many of the instruments have been satisfactorily established. To date, empirical information about construct and predictive validity are unavailable for instruments used in the state programs. Validities for the MCAT Skills Analysis tests range from .21 to .58 in relation to medical school basic science grades and National Board of Medical Examiners Test scores (Mitchell et al., 1986). Moderate correlations have been obtained between selected NBME scores and task-based supervisor ratings of job performance (Bryant, 1985).

WRITING SKILLS

The testing of writing ability is receiving increased attention at all educational levels. Many state and commercial testing agencies are currently involved in direct and indirect writing assessment programs. The National Assessment of Educational Progress (Mullis, 1985) also tests writing to reflect the differing purposes for which people write at home, at school, and in the community. The goals of these programs include establishing minimum competency, effecting remediation and instructional improvement, and/or informing selection, placement, and exit decisions. Many of the testing programs are designed to provide examinees with opportunities to demonstrate skill in such areas as: (1) developing a central idea; (2) synthesizing concepts and ideas; (3) separating relevant from irrelevant information; (4) developing alternative hypotheses; (5) presenting ideas cohesively and logically; and (6) writing clearly, observing the accepted practices of grammar, syntax, punctuation, and spelling.

Of the 33 state writing assessment programs, half rely exclusively on writing sample and the remainder have both a multiple-choice, indirect writing exam as well as a direct assessment of student writing. In addition to a student writing sample, the New Jersey High School Proficiency Test (Cooperman & Bloom, 1985), for instance, uses an objective format to test the ability to (1) select transition words to complete paragraphs, (2) identify usage errors in sentences and select appropriate corrections, (3) identify details inappropriate to main ideas, and (4) reorganize sets of sentences into logical order. The multiple-choice section of the New Jersey College Basic Skills Placement Test asks students to categorize ideas, use appropriate connectives, draw analogies, recognize principles of organization, use coordination and subordination appropriately, and use modifiers effectively (Barr & Hollander, 1980). The Connecticut Assessment of Educational Progress (Baron, 1984) uses a "revising test" whereby students are asked to proofread, edit, and revise writing samples with prespecified types of errors. They also use a dictation test whereby students are asked to transcribe and punctuate sentences with homonyms, different spellings, possessives, and possessive plurals. The California State University and Colleges English Placement Test has two,

35-minute multiple-choice sections called "Sentence Construction" and "Logic and Organization" (White & Thomas, 1981). To avoid overemphasis on the mechanics of writing, most state-testing officials and commercial publishers feel strongly that indirect assessment should be supplemented by generative demonstrations of writing ability. The Advanced Placement Test, administered by the Educational Testing Service, for example, uses both essays and multiple-choice exams to enable high school students to gain college credit. Reported correlations between direct and indirect assessments range from .30 to .75 across programs.

Several programs, like South Carolina's Basic Skills Education Entrance Examination (Meredith et al., 1985), the Texas Assessment of Basic Skills (Sachse, 1984), the Medical College Admission Test Essay Pilot Program (Mitchell, 1985), and the Maryland Functional Writing Program (Hermann & Williams, 1984) rely on one or two, 30- to 60-minute narrative, expository, or persuasive writing samples. A few use untimed exercise(s). The following is an example of an expository prompt from the MCAT Essay Pilot exam. A 45-minute response period is allotted for this writing task (Association of American Medical Colleges, 1985).

Loren Eiseley, in "The Unexpected Universe," describes a walk along a beach with the debris of life. Eiseley saw the sea rejecting its offspring; the life forms would fight their way home through the surf only to be cast back again upon the shore. As he walked, Eiseley came upon a stranger leaning over a starfish which was holding its body away from the stifling mud. With a quick yet gentle movement the stranger picked up the star and spun it out into the sea. "It may live," the thrower said, "If the offshore pull is strong enough."

Eiseley walked on, then looked back. For a moment in the changing light the thrower appeared to him magnified as though casting larger stars upon some greater sea. But Eiseley walked away from the star-thrower "in the hardened indifference of maturity." He was an observer and a scientist: he believed the star-thrower was crazy.

Later, after reflection, Eiseley returned to the beach. Silently, he sought and picked up a still-living star, spinning it far into the waves. He spoke once briefly. "I understand; call me another thrower." Then he thought, "The stranger is not alone any longer. After us there will be others."

Write an essay that explains what Eiseley was saying when he described himself as another thrower of starfish. Is being a thrower in conflict with being a scientist?

The following is a narrative prompt from the New Jersey High School Proficiency Test (Cooperman & Bloom, 1984). Students are given 30 minutes to respond.

Think of something important that happened in your life. It may have been happy or sad, painful or enjoyable. Write an essay in which you tell what happened and why it was important to you.

The English Composition Test, administered by the College Board, includes a single, 20-minute expository essay section (Kirrie, 1979). The following is a sample question.

First, consider carefully the following quotation. Then, read and follow the directions that are given in the assignment that follows the quotation.

"We have met the enemy and he is us."

Assignment: What does this quotation imply about human beings? Do you agree or disagree with its implications? Support you position with examples from your reading, observation, or experience.

The New Jersey College Basic Skills Placement Test also has a 20-minute essay section.

NAEP writing assessments include tasks which elicit informational, persuasive, and imaginative writing (Applebee et al., 1986). Informational writing is directed at presenting information and sharing ideas, for example, describing a trip, reporting on a science experiment, presenting political and social analyses. Persuasive writing attempts to influence the opinions or actions of others. Persuasive tasks call for the explication of a point of view and the provision of supporting details, for example, defending a morning or an afternoon school schedule. Imaginative writing prompts elicit personal or fictional narratives. For example, the 1984 assessment included an exercise based on a picture of a box with an eye peering through the opening. The prompt asked examinees to imagine themselves in the picture and to describe the scene and how they felt about what was going on around them.

Many of the state and several of the commercial programs use four- or six-point holistic rating scales to grade the essays. In these programs, trained readers rate the essays on their overall conformance to skill requirements outlined by the writing programs. Readers are trained to reach consensus on qualities relevant to grading. Papers are then generally read by two scorers with a third reading given to papers with insufficient score agreement. In most cases, holistic scoring is used by minimum competency or competency-based writing programs. Some programs use analytic or primary trait rating schemes whereby essays are scored on individual characteristics such as focus, organization, attention to audience, originality, sentence structure, word usage, and mechanics. Analytic and primary trait scoring methods provide useful information for diagnostic purposes. The National Assessment of Educational Progress uses primary trait scoring guides to isolate essential writing features and to present criteria for varied levels of performance on the features. A small number of programs, for example, South Carolina's Basic Skills Assessment and Education Entrance Examination Programs and the Texas Assessment Program use combined holistic and analytic or primary trait scoring schemes. The New Jersey High School Proficiency Testing Program uses a combined scheme called Registered Holistic Scoring which focuses on (1) organization/content, (2) usage, (3) sentence construction, and (4) mechanics (Braungart-Bloom, 1986). This method anchors writing features independent of topic and mode in an effort to ensure score point consistency from administration to administration.

Several of the state and commercial programs have come under fire for the inability to replicate results across essay prompts within and across modes of discourse and for the equivocality of standard setting procedures. Interrater reliability estimates of .65 to .80 have been derived for holistic scorings on the same topic across reader pairs. When variation in topics is introduced, however, the reliability of single-topic tests decreases to what is generally considered an unacceptable level. ACT's College Outcomes Measures Project reports a generalizability coefficient of .76 for a three-topic test. However, testing time constraints limit the number of topics that are administrable in most contexts. Research on prompt development and on the comparability of results across topics and modes of discourse is currently underway in a number of settings but few promising approaches to the development of topics similar in overall difficulty and prerequisite skills have been suggested. Research results on topic comparability within and across modes of discourse are mixed. Several investigators report that correlations between topics within a mode of discourse are higher than those across genres (Breland, 1986). Others report that topic effects obscure mode differences (Rock, 1986; Carlson et al., 1985).

The state writing programs generally use committees of experts to define minimum writing competency. The South Carolina Education Entrance Examination Program, for example, uses a modified Angoff multi-step procedure with its committee (Mappus et al., 1985). Standard setting procedures in several states have been criticized, however, either because unreasonably large percentages of students fail to test above the minimum competency levels, or because many upper level students fail to meet criterion at one or more administrations, or due to large discrepancies in the numbers of students scoring above criterion at different testings. The Maryland, South Carolina, and Texas programs have reported larger discrepancies in passing rates for equivalent groups on different topics. The Maryland Functional Writing Program's recent experience with large numbers of academically capable students scoring below the passing level on their essays is an example of the difficulty associated with setting valid competency standards. This program is now instituting revised criteria for functional writing which reward legible handwriting, good margins, and sentences with appropriate transition. New York, Nevada, and New Jersey also have writing competency criteria for graduation. South Carolina will operationalize graduation requirements in 1990.

The reliability problems noted earlier obviously limit the validity of direct assessments of writing ability. Some empirical information on the construct validity of writing assessments has been gleaned by the Illinois Inventory of Educational Progress (Chapman et al., 1984). This program reports correlations of .50 between essay data and inferential reading and complex multiple-choice grammar items and .45 between essay performance and complex mathematics problem solving. Correlations between the MCAT essay and text- and data-based problems are .43 and .38, respectively (Mitchell, 1985). Researchers for the City University of New York (CUNY) Freshman Skills Assessment report correlations ranging from .48 to .61 between their writing and reading comprehension tests (Ryzewic & Benchik, 1982). Correlations between the Test of English as a Foreign Language (TOEFL) experimental essay and the total operational TOEFL score, Law School Admission Test total, and Graduate Record Exam total are .72, .46, and .81, respectively (Carlson et al., 1985). Correlations between the essay portion of the College Board English Composition

\chievement Test and the Scholastic Aptitude Test Verbal and Mathematical sections are .56 and .35 (Breland & Jones, 1982). Factor analyses of analytic scorings on the English Composition Test essay identify two scorable aspects of writing samples, discourse skills, and mechanics skills. Correlations between these scores and self-reported writing ability range from .30 to .33 (Breland et al., 1984). The Medical College Admission Test Essay Pilot Program is now collecting validity data on their essay exam for currently enrolled medical students. They are also examining the incremental validity associated with writing assessments in medical school admissions.

LOGICAL AND ANALYTICAL REASONING

In October 1985, the Graduate Record Examination (GRE) introduced an operational test of logical and analytical reasoning (Leary, 1985; Wild, 1985). The items are designed to tap inference, deduction, and analysis skills prerequisite to graduate study. The logical reasoning items assess the abilities to understand, analyze, and evaluate arguments or parts of arguments. Tested skills include recognizing the point of an argument, recognizing assumptions on which an argument is based, drawing conclusions from given premises, inferring material missing from given passages, applying principles governing one argument or another, identifying methods of argument, evaluating arguments and counterarguments, and analyzing evidence. The analytical reasoning items require examinees to understand the structure of arbitrary relationships among persons, places, objects, or events to deduce new information from given relationships, and to assess the conditions used to establish the structure of relationships. Relationships may be based on temporal or spatial order, group membership, cause and effect, quantity, transitivity, reflexivity and/or conservation of quantity (Graduate Records Examination Board, 1986). Item 1 below is an example of a logical reasoning item; item 2 tests analytical reasoning (Graduate Records Examination Board, 1986).

1. If Ramon was born in New York State, then he is a citizen of the United States.
   The statement above can be deduced logically from which of the following statements?
   A. Everyone born in New York State is a citizen of the United States.
   B. Every citizen of the United States is a resident either of one of the states or of one of the territories.
   C. Some people born in New York State are citizens of the United States.
   D. Ramon was born either in New York or in Florida.
   E. Ramon is a citizen either of the United States or of the Dominican Republic.
2. In a certain society there are three marrage groups—J, G, and R.
   A man and a woman may marry if and only if they belong to the same group.
   The sons of group-J parents belong to group R; the daughters belong to group G.
   The sons of group-G parents belong to group J; the daughters belong to group R.

The sons of group-R parents belong to group G; the daughters belong to group J.

Which of the following statements can be inferred from the rules given?

I. A brother and sister may not marry.

II. A woman may not marry her father.

III. A man may not marry his granddaughter.

(A) I only (B) II only (C) III only
(D) I and II only (E) I, II, and III

Efforts to refine and extend the constructs assessed by the GRE reasoning test are currently in progress. McPeek, Tucker, and Chalifour (1985) are collecting additional data on the analysis and reasoning processes deemed essential to successful graduate study. Skills such as: (1) integrating partial conceptualizations into a broader view; (2) formulating relevant concepts from disparate data and facts; and (3) formulating alternative conceptualizations of explanations or conclusions are currently under consideration.

In 1984, the State of Florida Department of Education defined performance standards in reading, writing, and computation for the award of associate of arts degrees from community colleges and state universities and for admission to upper division status in state universities (College Level Academic Skills Project, 1985). Logical reasoning items play a major role in the computation skill area. Reasoning items focus on the abilities to recognize invalid arguments leading to valid conclusions, determine equivalence and nonequivalence of statements, infer valid reasoning patterns, and draw logical conclusions from facts and data (College Level Academic Skills Project, 1986). In 1983, the Totowa Board of Education and a task force from the New Jersey Department of Education developed the New Jersey Test of Reasoning Skills (Shipman, 1983). This test, intended for students in grades four through college, addresses classical syllogism, the meaning of categorical statements, the identification of assumptions, and induction. The test is not administered on a statewide basis. The New Jersey Test of Reasoning Skills and the Florida logical reasoning items are similar in format to the GRE reasoning items.

Research on the psychometric characteristics and construct validity of the GRE reasoning tests has yielded reliability estimates of .90 and correlations between GRE reasoning and verbal and GRE reasoning and quantitative items from .64 to .73 (Kingston, 1984). Regression analyses reveal that the GRE reasoning score contributes a small amount of unique information to the prediction of undergraduate grade-point averages. Factor analytic examinations of the predictor space for the GRE suggest a unique reasoning factor.

Predictive validity data on the GRE reasoning tests for performance in graduate school are inconclusive. Kingston (1984) found that the reasoning measure contributed little to the predictive validity of the battery in relation to first-year grade-point average in graduate school. There was some evidence of incremental validity for foreign and minority students, however. Local validity studies provide some support for increases in the predictability of advancement to doctoral candidacy decisions (Mowsesian & Hayes, 1985). Data on the predictive validity of the Florida logical reasoning items and the New Jersey Test of Reasoning Skills are unavailable.

NATIONAL ASSESSMENT OF EDUCATIONAL PROGRESS

The National Assessment of Educational Progress (NAEP) is an ongoing, congressionally mandated project designed to assess "the performance of children and young adults in the basic skills of reading, mathematics and communication and to report data periodically on changes in the knowledge and skills of such students over a period of time" (General Education Provision Act, 1968, 1978; Goodison, 1985). Assessments in art, career and occupational development, citizenship, literature, music, science, social studies, and writing have also been conducted by NAEP. Data collection generally focuses on 9-, 13-, and 17-year olds.

In 1985, NAEP completed an interesting supplementary study of adult literacy (Mullis, 1985; NAEP, 1985). There were 3,600 21- to 25-year olds who participated in the assessment. Literacy was defined by NAEP as the use of printed matter in a variety of contexts to function effectively in society, achieve one's goals, and develop one's knowledge and potential. NAEP officials stated that the assessment documented people's strategies for interacting with written material and related this interaction to performance and practice. The assessment addressed the everyday, information handling capabilities of young adults. The content domain for literacy assessment was defined by panels of experts. The assessment employed a wide range of hands-on tasks involving the interpretation and discussion of narrative and tabular material. Stimuli included, for example, grocery ads, newspaper text, the Yellow Pages, and the World Almanac. Examinees were asked to locate data, interpret and manipulate information, and present results. In one of the simpler narrative tasks, for instance, the respondent was presented with a series of six photographs showing a sequence of events—a young man waking up ill, going to a doctor's office, seeing a doctor, getting a presciption, and taking some medicine. The individual was asked to tell the story depicted by the photographs (Mead & Campbell, 1981). The task required narrating a sequence of events. The practical nature of the information processing requirements of the assessment tasks contributed to the face validity of the testing.

Assessment data suggested that relatively few adults reach levels of information processing proficiency associated with complex and demanding tasks. Most of those surveyed were proficient at tasks at the lower end of the skill continuum and more than half were rated at moderate levels of competency. Because the cohort of 18- to 23-year olds in the United States is expected to decrease over the next decade and because the demographic composition of the group is projected in such a way that the subpopulation of young adults performing at lower levels will disproportionately increase, NAEP officials recommend that action be taken to improve the literacy skills of young people. NAEP's adult literacy results provide important information about characteristics of the human resource pool which will be available to public and private industry and higher education in future years.

SUMMARY

This chapter has looked at efforts to expand the testing domain by federal, state, and commercial large-scale achievement testing programs. The domain specifications, assessment instruments, and evaluation plans for critical thinking skills and problem solving, writing ability, logical and analytical reasoning, and adult literacy programs have been discussed. The research and development activities that underly these testing programs are generally quite impressive. The care involved in drafting content specifications for many of the assessment projects is evident. Most of these programs were conceptualized and instituted after many states introduced basic skills and minimum competency testing legislation in the late 1970's and early 1980's; evaluative data are, therefore, as yet largely unavailable. Construct validity data have been published for the GRE analytical test and for some of the writing programs and they point to the measurement of unique abilities. Several programs are currently collecting predictive validity data for their measures. Data on the predictive validity of the MCAT Skills Analysis and NBME tests are promising. Information on the incremental validity of the GRE reasoning items also is available.

REFERENCES

Applebee, A. N., Langer, J. A., and Mullis, I. V. S. (1986). *NAEP writing trends across the decade, 1974-1984.* Paper presented at the annual meeting of the American Educational Research Association, San Francisco, CA.

Association of American Medical Colleges. (1984). *The MCAT student manual.* Washington, DC: Author.

Association of American Medical College. (1985). *The MCAT essay test booklet.* Washington DC: Author.

Baron, J. B. (1984). Writing assessment in Connecticut: A holistic eye toward identification and an analytic eye toward instruction. *Educational Measurement, 3* (1), pp. 27-28.

Barr, E. E., and Hollander, T. E. (1980). *New Jersey College Basic Skills Placement Test: Scoring the essays.* Trenton, NJ: New Jersey State Department of Education.

Braungart-Bloom, D. S. (1986). *Assessing higher order thinking skills through writing.* Paper presented at the annual meeting of the American Educational Research Association, San Francisco, CA.

Breland, H. M., and Jones, R. J. (1982). *Perceptions of writing skill.* (College Board Report No. 82-4). New York: College Entrance Examination Board.

Breland, H. M., Rock, D. A., Grandy, J., and Young, J. W. (1984). *Linear models of writing assessments* (Research Memorandum 84-2). Princeton, NJ: Educational Testing Service.

Bryant, S. K. (1985). Report of the Board of Trustees meeting. *NBRC Newsletter, 11* (1), p. 1.

Carlson, S. B., Bridgeman, B., Camp, R., and Waanders, J. (1985). *Relationship of admission test scores to writing performance of native and non-native speakers of English* (TOEFL Research Report 19). Princeton, NJ: Educational Testing Service.

Chapman, C. W., Fyans, L. J., and Kerins, C. T. (1984). Writing assessment in Illinois. *Educational Measurement, 3* (1), 24–26.

College Level Academic Skills Project. (1985). *CLAST test administration plan 1985-1986.* Tallahassee, FL: State of Florida Department of Education.

_____ . (1986). *CLAST sample test items.* Tallahassee, FL: State of Florida Department of Education.

Comer, A. (1986). Test service admits to surprise section. *The Cavalier Daily, 97* (27), p. 1.

Cooperman, S., and Bloom, J. (1984). *New Jersey statewide testing system: High school proficiency test, 1983-1984.* Trenton, NJ: New Jersey State Department of Education.

_____ . (1985). *High school proficiency test skill array: Writing.* Trenton, NJ: New Jersey State Department of Education.

Education Commission of the States. (1982). *The information society: Are high school graduates ready?* Denver, CO: Author.

_____ . (1984). *Current states of state assessment programs.* Denver, CO: Author.

Ennis, R. H. (1985). *A logical approach to measuring critical thinking skills in the fourth grade.* Paper presented at the annual meeting of the American Educational Research Association, Chicago, IL.

Florida Department of Education. (1984). *Student performance standards of excellence for Florida schools in mathematics, science, social studies and writing.* Tallahassee, FL: Author.

General Education Provision Act. (1968, 1978). Title IV of PL 90–247, as amended by PL 95–561, Section 405(K)(1).

Goodison, J. (1985). *National Assessment of Educational Progress: An update of the data collection process.* Paper presented at the annual meeting of the American Educational Research Association, Chicago, IL.

Graduate Records Examinations Board. (1986). *GRE information bulletin 1986-1987.* Princeton, NJ: Educational Testing Service.

Hermann, F., and Williams, P. L. (1984). Writing assessment in Maryland. *Educational Measurement 3* (1), 23–24.

Hixon, S. J., and Loadman, W. E. (1986). *The dimensionality of measures derived from a nationally standardized written clinical simulation examination.* Paper presented at the annual meeting of the American Educational Research Association, San Francisco, CA.

Kingston, N. (1984). *Reanalysis of the psychometric characteristics of the revised analytical measure of the GRE general test.* Educational Testing Service, New Jersey; Unpublished manuscript.

Kirrie, M. (1979). *The English composition test with essay.* Princeton, NJ: The College Board.

Kneedler, P. E. (1985). *Assessment of critical thinking skills in history-social science.* Sacramento, CA: California State Department of Education.

Leary, L. F. (1985). *The next step: Directions for future research.* Paper presented at the annual meeting of the American Educational Research Association, Chicago, IL.

Mappus, L. L., Sauders, J. C., and Meredith, V. H. (1985). *Scoring criteria and standard setting.* Paper presented at the annual meeting of the American Educational Research Association, Chicago, IL.

McPeek, W. M., Tucker, C., and Chalifour, C. (1985). *The analytical score: What it measures and what it doesn't.* Paper presented at the annual meeting of the American Educational Research Association, Chicago, IL.

Mead, N. A., and Campbell, A. (1986). *Profiles for literacy: Oral-language refining.* Paper presented at the annual meeting of the American Educational Research Association, San Francisco, CA.

Meredith, V. H., Brinlee, P. A., and Mappus, L. L. (1985). *Prompt development and the prequating procedures used for prompt selection.* Paper presented at the annual meeting of the American Educational Research Association, Chicago, IL.

Mitchell, K. J. (1985). *Preliminary results from the MCAT essay pilot project.* Association of American Medical Colleges. Unpublished manuscript.

Mitchell, K. J., Anderson, J., and Beran, B. (1986). Vital statistics on the Medical College Admission Test. *The Advisor, 6* (2), 25–31.

Mowsesian, R., and Hayes, W. L. (1985). *Comparative validity of the GRE-Analytical Test.* Paper presented at the annual meeting of the American Educational Research Association, Chicago, IL.

Mullis, I. V. S. (1985). *NAEP perspectives on literacy: A preview of 1983–1984 writing assessment results, the young adult literacy assessment and plans for 1986.* Paper presented at the annual meeting of the American Educational Research Association, Chicago, IL.

National Assessment of Educational Progress. (1985). Adult literacy probe moves forward: *NAEP Newsletter,* pp. 1f.

National Science Board Commission. (1983). *Educating Americans for the 21st century: A plan for action improvising mathematics, science and technology education for all American elementary and secondary students so that their achievement is ths best in the world by 1985.* Washington, DC: National Science Board Commission on Precollegiate Education in Mathematics, Science and Technology, National Science Foundation.

Rock, D. A. (1986). *Essay tests: What is being measured, how does it differ from what is being measured by multiple-choice tests.* Paper presented at the annual meeting of the American Educational Research Association, San Francisco, CA.

Ryzewic, S. R., and Benchik, G. (1982). *The CUNY Writing Assessment Test: A three-year audit review 1979–1981* (Research Monograph Series No. 2). New York: The City University of New York.

Sachse, P. P. (1984). Writing practices in Texas: Practices and problems. *Educational Measurement, 3* (1), pp. 21–23.

Shipman, V. (1983). *New Jersey test of reasoning skills.* Totowa, NJ: Totowa Board of Education.

White, E. M., and Thomas, L. L. (1981). Racial minorities and writing skills assessment on the California State University and colleges. *College English, 42,* 276–283.

Wild, C. L. (1985). *The analytical score: Research leading to a new measure.* Paper presented at the annual meeting of the American Educational Research Association, Chicago, IL.

7 Cognitive Modeling of Learning Abilities: A Status Report of LAMP

Patrick C. Kyllonen
Raymond E. Christal

INTRODUCTION

Considerable headway has been made during the last decade in our understanding of human cognition. This has led to speculation that it is only a matter of time before an improved technology for gauging individuals' intellectual proficiencies will be developed. The stakes are high: psychological testing of cognitive proficiency is presently widespread in industry, the schools, and the military. Improved tests would have a profound economic impact in cutting education and training costs and enabling a more efficient and fair system of personnel utilization. Although the concept of psychological testing must certainly be considered one of psychology's true success stories, it is also primarily a past accomplishment. Systematic studies of predictive validity have shown that today's aptitude tests are no better than those available shortly after World War II (Christal, 1981; Kyllonen, 1986).

Support was provided by the Air Force Human Resources Laboratory and the Air Force Office of Scientific Research, through Universal Energy Systems, under Contract No. F41689-84-D-0002/58420360, Subcontract No. S-744-031-001. We thank Valerie Shute, Bill Tirre, and William Alley for their comments on this report, and we give a special acknowledgement to Dan Woltz for many long and thorough discussions of the issue addressed herein.

But even if it is agreed that forces are conspiring to usher in a new era of cognitive testing, there still is considerable debate on exactly what form these new cognitive tests will take. On one side of the debate, some argue that what cognitive psychology has to offer is a rationale and a methodology for measuring basic information processing components (Detterman, 1986; Jensen, 1982; Posner & McLeod, 1982). According to this view, the cognitive test battery of the future would consist of measures of speed of retrieval from long-term memory, short-term memory scanning rate, probability of transfer from short- to long-term store, and the like. On the opposite end of the debate are those who suggest that the fundamental insight of cognitive science is that cognitive skill reflects primariliy knowledge rather than general processing capabilities. This perspective has led to calls for testing intermingled with instruction, testing aimed at measuring what students know and what they have learned in the context of their current instructional experience (Embretson, in press; Glaser, 1985). This has been called *steering testing* (Lesgold, Bonar, & Ivill, 1987) or - *apprenticeship testing* (Collins, 1986). Between these positions are those who propose new kinds of cognitive tests that are not radically different from existing ones, but perhaps richer and more diverse in what they measure (Sternberg, 1981b); Hunt & Pellegrino, 1984).

In this chapter, we provide a status report of one ongoing program of research, the Learning Abilities Measurement Program (LAMP), that has been concerned with developing new methods for measuring cognitive abilities. We discuss some of our early thinking on the implications of cognitive psychology for testing, and how we have adjusted our ideas in light of data and further reflection. We conclude with a brief discussion of CoLoSsal, the Complex Learning Skills Laboratory, the setting in which we hope to validate the new tests.

COGNITIVE THEORY AND APTITUDE TESTING

The idea of grounding psychological testing in cognitive theory is not entirely novel. During the 1970s and 1980s, the Air Force Office of Scientific Research (AFOSR) and especially, the Office of Naval Research (ONR), supported a number of basic research projects which had the explanation of individual differences in learning and cognition as a central goal. This research largely concentrated on the analysis of conventional aptitude tests, probably for two reasons. First, analysis of aptitude tests is important in its own right, as an attempt to determine what it is that such tests measure. But, second, and perhaps more importantly, aptitude tests can be viewed as generic surrogates for tasks tapping more complex, slowly developing learning skills. It is difficult and extremely expensive to identify and analyze the information processing components associated with the acquisition of computer programming skill—so goes the argument—it is far cheaper and more efficient to analyze the seemingly more tractable components of some aptitude test, such as an analogies test, that predicts success in computer programming. And the fact that tests do such a good job in predicting training outcomes can be taken as evidence that pretty much the same cognitive components are involved in both test taking and learning.

The wave of aptitude research that was motivated by these considerations did not lead directly to improvements in existing aptitude testing systems, however. A number of new methods and techniques, such as cognitive correlates analysis (Hunt, Frost, & Lunneborg, 1973) and componential analysis (Sternberg,1977) were developed for analyzing aptitude tests, but the application of these methods did not suggest how the tests themselves might be improved. There have been suggestions that cognitive tasks exported from the experimental psychologist's laboratory might somehow be used to supplement or even replace existing aptitude tests (Carrol, 1981; Hunt, 1982; Hunt & Pellegrino, 1984; Pellegrino & Glaser, 1979; Rose & Fernandez, 1977, Snow, 1979; Sternberg,1981b), but after almost ten years, the research still has not yet been carried out to an extent sufficient for determining whether this is really feasible.

Probably the reason cognitive-based aptitude research has not translated already into better tests is that this has not been a primary goal of the research. Indeed, if the creation of better tests had been the primary goal, the approach of analyzing and decomposing existing tests does not seem very promising. If such research efforts were completely successful, "if the research turned out better than anyone's wildest expectations," at best new tests would simply duplicate the validity of existing tests.

LEARNING ABILITIES MEASUREMENT PROGRAM (LAMP)

In contrast to some of the aptitude research projects just discussed, our own work in connection with Project LAMP, has from its inception been focused on the goal of developing an improved selection and classification system. Our current efforts fall into two categories. First, we are continuing to model basic cognitive learning skills and their interrelationships, and to explore different methods for measuring these skills. Second, we have more recently begun thinking seriously about a system for validating the new cognitive measures. The system involves the extraction of learning indexes both on short-term (one hour) and long-term (one week) learning tasks, that will serve as criteria against which the new cognitive measures will be validated. Although we have not yet collected data on the long-term learning tasks, we have set up the laboratory, which consists of 30 computerized tutoring systems. In the remainder of this chapter, we will discuss these two categories of ongoing LAMP research. We begin with a discussion of studies that have attempted to measure cognitive skills.

Modeling Cognitive Skills: The Four-Source Framework

Much of our work on identifying basic learning skills has centered around what we have called the four-source framework (Kyllonen, 1986). This is the idea that individual differences in a wide variety of learning and performance tasks are due to differences in four underlying sources: (a) effective cognitive *processing speed*, (b) effective *processing capacity*; and the general breadth, accessibility, and pattern of one's (c) conceptual *knowledge* and (d) procedural and strategic *skill*. Figure 7.1 illustrates these relationships.

We refer to the knowledge and skill components of this model (components [c] and [d]) as *enablers*, in the sense that any learning or performance task can be characterized as consisting of a necessary set of knowledge and skill prerequisites. We refer to the processing speed and working memory components of the model ([a] and [b]) as *mediators*, in the sense that these components mediate the degree to which the learner or problem-solver is able to use his or her knowledge and skills effectively. We have found the four-source framework to be useful in organizing our own as well as others' research and in monitoring our research progress. Further, although we have not yet applied it widely in this fashion, we expect that the system will be useful for task analysis purposes.

Thus far, most of the research we have accomplished in connection with the four-source proposal has been concerned with (1) improving the way in which we measure cognitive skills and (2) determining the dimensionality of the skills and subskills embedded within the four-source model. We now turn to a discussion of the four components, in turn.

Processing Speed

Considerable research on individual differences in cognition over the past ten years has been concerned with determining the relationship between processing speed and performance on complex tasks, such as intelligence tests. There are a number of reasons for the high level of interest in processing speed. One reason is that we now can measure it. The availability of microcomputers as testing instruments makes it feasible to measure response time to particular items with precision. Paper and pencil tests only allowed gross estimates of response speed. Two, processing speed seems to reflect something basic, something fundamentally a part of all mental activity, and therefore something that might explain the general factor in intelligence, in some sense. Third, processing speed has played a major role in cognitive theories in revealing the dynamics of mental processes since the beginnings of modern cognitive psychology. Neisser's (1967) book, which is generally considered the kickoff point for the discipline, reported primarily on reaction time studies. Finally, there are operational performance contexts, such as the air traffic controller workstation or the cockpit, that require efficient processing of considerable data. Understanding the relationship

Figure 7.1
Four-Source Research Framework

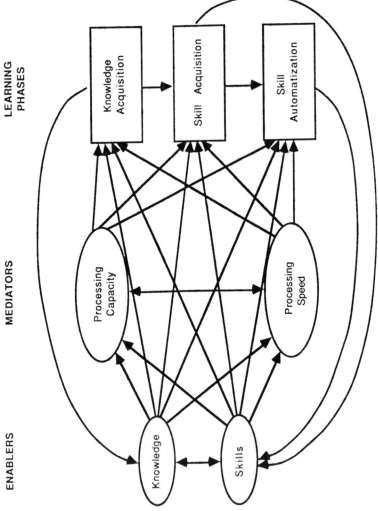

between processing speed and performance in these kinds of contexts would have immediate practical payoff.

In our own laboratory, we have conducted a number of studies on processing speed that have focused on both its psychometric properties and its relationship to performance on criterion tasks. Studies have run the gamut in addressing both applied and basic issues. A number of early studies in the project (reported in Kyllonen, 1985) were designed simply to address the question of whether processing speed could be more appropiately characterized as a unitary or multidimensional construct. That is, we addressed the question of whether some people are generally faster information processors than others, or whether it is more appropriate to think in terms of varieties of processing speed. Both positions can be argued for on rational grounds. Much of Jensen's work (Jensen, 1982) at least implicitly presumes a general speed factor. But low correlations between processing speed tasks and measures of general intelligence have led others to propose multiple, correlated processing speed components (for example, Detterman, 1980).

One way to address the dimensionality question is simply to measure response time on a wide variety of cognitive tests, such as those one finds in the ETS kit, and perform a factor analysis on the resulting scores. In one study (Kyllonen, Tirre & Christal, 1985), we did just that, and found evidence for both separate reasoning, quantitative, and verbal processing factors, and a higher order general processing speed factor. Interestingly, we found that although processing speed scores were quite reliable, at least within session, they were not related to accuracy scores on the same tests. Timed versions of the tests thus mix these two separable components of performance in yielding only a single score. There are problems with this approach to testing the dimensionality question, such as how to allow for speed-accuracy trade-off. what to do with response times when the person guessed incorrectly, and so forth. But a more substantive problem is that although the findings are suggestive, they fall considerably short of revealing much about the process that produced them.

Thus in subsequent work we have restricted our focus (and employed a narrower range of tasks) in the hopes of achieving a better process-oriented understanding of the generality question. In these studies, we attempt to identify processing stages, then measure the duration of those stages for individual subjects, then compute the stage intercorrelations. The procedure is best illustrated by example. In the first study (Kyllonen, 1987), we administered a series of tasks that required subjects simply to determine whether two words presented (for example, *happy lose*) were similar or dissimilar with respect to valence. *Happy* would be considered a positive valence word, *lose* would be considered a negative valence word. We presumed that a decision on this task was executed after a series of processing stages. The subject begins by *encoding* one of the words, then encoding the second word. The result of the encoding process is that a symbol representing valence is deposited in working memory for each word. The subject then *compares* those symbols. The result of the comparison process is an implicit assertion that the symbols are either the same or different. A *decision* process then takes the comparison result and translates it into a plan for the execution of the motor response. A *response* process then executes the motor response. Through the method of precuing, which has been used with some success in separating process components on other reaction time tasks (for example, Sternberg, 1977), we were able to independently estimate the duration of each of these processing stages.

We also administered two other versions of the task in which the only difference was that subjects were required to (1) decide whether two digits were the same with respect to oddness or evenness, or (2) decide whether two letters were the same with respect to vowelness versus consonantness. The data analysis addressed two questions regarding generality. First, were parallel measures of stage duration (estimates derived from separate blocks of items) more highly intercorrelated than correlated with other stage durations? This is a direct test of stage independence. Second, were stage durations estimated from tasks with different content (words versus digits versus letters) more highly intercorrelated or were alternative stages taken from same-content tasks more highly intercorrelated? This is a direct test of the relative importance of content versus process. Although the analyses are rather complex, the general finding was that processes were somewhat independent, and also general across contents. That is, fast encoders are not necessarily fast comparers, but fast encoders on the word task were also fast encoders on the digit task.

One of the problems with this approach to studying dimensionality is that it relies on a model of performance that assumes serial execution of processing stages. In our more recent work (Kyllonen, Tirre, & Christal, 1987) we have relaxed this assumption by applying both models that assume serial execution and those that do not, in estimating stage durations. (We also have abandoned the precuing technique because its validity depends on the serial execution assumption.) Following Donaldson's (1983) analysis, stage durations can be estimated in two ways. Assume an ordered set of tasks, each of which can be characterized as requiring a proper superset of the processes of its predecessor. For example, the following set of tasks, each of which requires processing a pair of words, might be characterized this way: reaction time, choice reaction time, physical matching, name matching, and semantic (meaning) matching. That is, reaction time consists only of a *reaction* component, the choice task adds a *decision* component, the physical matching task adds *comparison*, name matching adds *retrieval* from long-term memory, and semantic matching adds *search* through long-term memory.

One can estimate each of these stage durations either by subtracting latency on the predecessor task from latency on the target task (the difference score model), or by statisically holding constant the duration of all predecessor tasks (the part correlation model). The two models make different assumptions about the relationships among task components. The difference score model assumes nothing about the relationship between the duration of the target component (for example, comparison) and the duration of the predecessor task (for example, choice reaction time). Thus this correlation is a parameter to be estimated. But the cost of this flexibility is the assumption that the duration of the target component (for example, comparison) is the same, regardless of whether the component is embedded in the physical matching task, the name matching task, or whatever. Conceptually there are two problems with this assumption. Consider the reaction component. It may be that reaction is fast when nothing else is going on, as on the simple reaction time task, but slow when it follows complex processing, as on the semantic matching task. Or it could be the opposite, due to parallel processing: reaction appears slow on the simple reaction time task because it is the only process executing; but on the meaning identity task, the reaction begins before decision ends, and thus appears fast (as is specified in process cascading models, McClelland, 1979).

The part correlation model avoids this assumption and allows for variability in stage durations over different tasks. This is represented as freedom in the regression weight associated with stage duration to differ from one. But in order to achieve this flexibility, the part correlation model must compensate with an assumption not required with the difference score model. In the part correlation model, it is assumed that the duration of the target stage is uncorrelated with the duration of the predecessor task. For example, the duration of the comparison component in the context of the physical matching task would be assumed to be uncorrelated with response time on the choice reaction time task.

Which of these sets of assumptions is correct, those associated with the part correlation model, or those associated with the difference model? It is not possible to tell, but it is possible to employ both models and then to be confident of relationships only when the models agree.

We took this approach in attempting to estimate the relationship between processing stage durations and performance on a vocabulary test, and also on a paired associates learning task. Vocabulary is an interesting test case because it is a good measure of general intelligence. The current view is that breadth of word knowledge reflects efficient learning processes in inferring word meanings in context (Marshalek, 1981; Sterberg & Powell, 1983). An additional motivation for looking at vocabulary as a criterion was that a considerable literature has evolved from Hunt and colleagues' (Hunt, Frost, & Lunneborg, 1973) early finding of a relationship between the duration of the retrieval stage (as estimated by the difference between response time on the name and physical matching tasks) and verbal ability.

Contrary to Hunt et al. and other previous work, however, we did not find much of a relationship between *retrieval* speed and vocabulary ($r = .17$, $N = 710$). But we did find a strong relationship between *search* speed and vocabulary ($r = .49$). Subjects capable of quickly accessing semantic attributes of words, controlling for how quickly they did other kinds of information processing, had greater vocabularies than did other subjects.

We found a similar relationship between processing speed and learning, but only in particular circumstances—namely, when study time on the learning task was extremely short (.5 to 2 seconds per pair). The component analysis again made it possible to isolate the semantic search component, as opposed to other processing speed components, as the one consistently most critical in determining learning success. Over a number of studies (which varied on block size, recognition versus recall responses, etc.), the correlation between learning success and response time on the meaning identity test, controlling for (or eliminating by subtraction) response time on other information processing tests ranged from $r = .30$ to $r = .50$. In some studies, other information processing speed components predicted learning outcomes, but only inconsistently.

We currently are engaged in two lines of extension to the processing speed work. One is motivated by the idea that information processing speed may be closely tied to working memory capacity insofar as both measures reflect the dynamic activation level of a memory trace (Woltz, 1988). An intriguing implication of this idea has to do with individual differences in the maintenance of activation. In most learning tasks, we do not simply access a term once and only once. Rather there is redundancy in instructional materials, which allows for multiple accesses of a concept in an instructional episode. Thus the important search speed variable is not just how quickly a concept can be accessed on first encounter, but also how quickly the concept can be reaccessed on second, third, and fourth encounters. Woltz (1988) has shown not only that subsequent accesses are much faster than first encounters, but that there is substantial individual differences in the amount of improvement in speed from first to subsequent encounters. Interestingly, those who benefit most are not necessarily those who are quickest to begin with. We explore further ramifications of the idea of activation as a concept underlying working memory capacity in the next section.

A second extension to the processing speed work involves the exploration of reaction time distributions as a way of determining how subjects process items. There is some work (Hockley, 1984; Ratcliff & Murdock, 1976) suggesting that reaction time on simple tasks actually reflects two underlying components: a normally distributed processing component (for example, true comparison time) and an exponentially distributed waiting time component (for example, time of attention lapses and the like). We are currently investigating the feasibility of estimating these reaction time components and determining whether they reflect reliability processes (Fairbank, 1988).

In summary, we have and are continuing to explore a number of mathematical models for identifying component processing speed, and for determining the relationship between different kinds of processing. One benefit from this kind of analysis is that it enables the determination of whether processing speed is a single construct or whether there are multiple varieties of processing speed (the latter appears to be the case). The implication for test development has to do with how, and how many different kinds of tests will be necessary to measure processing speed.

A second benefit from this kind of analysis is that it allows one to determine what kind of cognitive processing affects learning (in different contexts). One result is that it appears that general reaction speed is not highly related and therefore fundamental to learning as might be expected on the basis of work by Jensen and others. We have found relationships between basic reaction time and learning, but the particular component of speed of searching semantic memory appears to be the more critical predictor of verbal learning success. This is shown both in studies employing vocabulary scores as a criterion and in those employing a highly speed presenatation of material to be learned. (Perhaps both tasks reflect the learner's ability to quickly elaborate on the stimulus material.)

Processing Capacity

Although much of the early work on the project was concerned with response time, we recently have begun focusing more attention on similar kinds of analyses of working memory capacity. It now appears, not only on the basis of our own work (Woltz & Christal, 1985), but on work from a number of laboratories (Anderson & Jeffries, 1985; Daneman & Carpenter, 1980; Hitch, 1978), that this component of the information processing system is responsible for learner differences on a wide variety of learning tasks.

In keeping with contemporary views of the human cognitive architecture, we propose that working memory may be defined as that portion of memory currently in a highly active or accessible state, that is, whatever is being processed or attended to at any given time. The individual differences corollary is that greater working memory capacity should be associated with greater attentional and learning capabilities. Woltz (1988) has pointed out that this quite general description of working memory is realized in the literature in two rather different forms, which we will refer to as the *processing workspace* and *activation capacity* models.

The *processing workspace* model of working memory, due largely to the work of Baddeley and Hitch (1974), proposes a limited, consciously-controlled, short-term memory capable of storing roughly three to nine items at a time. The capacity of this structure is determined mostly by how efficiently one processes new incoming information. Much of our work on working memory to date has been the application of the processing workspace model to the development of working memory capacity tasks. The guiding construction principle is that the task requires the retention of some information, while simultaneously requiring the processing or transforming of other information. This principle is consistent with Baddeley and Hitch's (1974) original definition, and seems on the surface to lend itself readily to ecologically valid tests of memory capacity insofar as much of learning demands simultaneous retention and processing. In contrast, what is required on span tests seems contrived and not typical of what people actually do when engaged in realistic learning.

Figure 7.2 shows sample items from various tests developed in our laboratory as well as a couple of borrowed tests we have administered. In the first test, "ABCD Order," the subject is informed that all items involve two sets of letters. The first set is defined as the letters A and B, and the second set as the letters C and D. The subject is then presented a series of trials consisting of three statements that constrain the ordering of the four letters. In the item pictured, for example, the subject is presented a frame (the task is presented by computer, one frame at a time), which states that Set 1 follows Set 2. The subject is expected at this point to note that the letters A and B will follow the letters C and D in the final list. On the second frame, the subject is informed that B precedes A. On the third frame, the subject is told that D is not followed by C (that is, D follows C). The frames are presented successively, and the subject cannot look back to retrieve previous statements. From the three assertions, the subject would be expected to generate the proper ordering of the four letters, CDBA. The test probe is then presented, and the subject would respond according to whether it matched her memory string.

Another test, "ABC Assignment" also involves successive presentation of instruction frames, only here the instruction frames are assignments of either values (for example, A = 3), expressions (for example, A = 24 -17), or equations (for example, A = B / C). In the item pictured, the subject first sees that A gets the value of B divided by 2. The subject does not yet know what B is and so must remember the equation. The next frame states that C gets the value of B minus A. Again, the subject still does not know the value of B and so must remember the equation. Finally, the subject is shown that B is 6 and this allows him to solve for C and A, But in order to do so he must remember the equations for C and A. The subject is then tested for which values he can remember.

In the third test, "Alphabet Transformations," the subject is shown either one, two, or three random letters, then on the next frame is instructed to either add or subtract one, two, or three (n). Add and subtract in this context means to determine which letter precedes or follows each of the target letters by n positions. The other tests in Figure 7.2, the AB Verification test and the Sunday-Tuesday test are probably self-explanatory. They are borrowed from Baddeley (1968) and Hunt, Frost, and Lunneborg (1973), respectively.

As with information processing speed, an important initial question to be asked regarding performance on these kinds of tasks is whether working memory capacity is a unitary or multidimensional construct. A related question concerns the relationship between working memory and performance on other more conventional aptitude tests. We addressed both questions in

Sample Test Items Measuring Working Memory Capacity

Test Name	Frame 1	Frame 2	Frame 3	Test Probe 1	Test Probe 2	Test Probe 3
"ABCD Order"	Set 1 follows Set 2	B precedes A	D is not followed by C	"CDBA" T or F?		
"ABC Assignment"	A = B/2	C = B-A	B = 6	A = ?	B = ?	C = ?
"Alphabetical Transformation"	"CFM"	Subtract 3		"EHO" T or F?		
"AB Verification"	A precedes B			BA T or F?		
"Sunday-Tuesday"	Sunday + Tuesday			= Wednesday T or F?		

157

a large-scale correlational study recently completed (Christal, 1987). We administered the tests shown in Figure 7.2, along with additional measures such as Memory Span and Complex Arithmetic (mental arithmetic where calculations requires holding intermediate results, for example, 453-296; 132 / 3; 14 × 12). Additionally we had available subjects' scores on the Armed Services Vocational Aptitude Battery ASVAB, which consists of 10 paper and pencil tests, such as Word Knowledge, Numerical Operations (Number Facts), Auto and Shop Knowledge, and General Science Information.

A correlation matrix was generated from the percent correct scores on the computerized tests and the raw scores on the timed ASVAB tests. A principal factor analysis of this matrix yielded four factors. A Working Memory factor was defined primarily by ABC Assignment ($r = .80$), but also was heavily loaded by ABCD Order, Complex Arithmetic, and the other working memory measures (all of which showed $r > .60$). The two verbal measures, Word Knowledge and Paragraph Comprehension, had only modest loadings on this factor ($r < .10$). In addition to the Working Memory factor, separate Verbal and Speeded-Quantitative factors were extracted. The Verbal factor was defined by Word Knowledge ($r = .77$), but also was significantly loaded by both the ABCD Order task, and AB Sentence Verification, which may be thought of as an abridged version of the ABCD test ($r > .50$). The Speeded-Quantitative factor was defined by the Numerical Operations test ($r = .75$) but it also was significantly loaded by the Complex Arithmetic and the Sunday-Tuesday tests ($r > .30$). The basic pattern of results found here have been corroborated in a recently completed follow-up study.

Taken together, the results suggest the involvment of both domain knowledge (quantitative and verbal) and a domain-independent working memory in memory test performance. In addition, it appears from the data over the two studies that the working memory factor subsumes the reasoning factor. That is, individual differences in reasoning proficiency may be due entirely to differences in working memory capacity. Christal notes that the factor on which all the reasoning tests in the battery loaded highly is a working memory factor in that the test defined it, Alphabet Transformations ($r = .68$, in the follow-up study), does not appear to involve reasoning per se, but does clearly depend on working memory capacity.

Recently we have begun investigating an alternative to the processing workspace model, which is based on a very different conceptualization of working memory. The activation capacity model, based primarily on Anderson's (1987) ACT* theory, defines working memory not as a separate short-term store, but rather as a state of fluctuating activation patterns characterizing traces in long-term memory. According to this theory, long-term memory is a network of traces, each characterized by resting activation levels. Traces become activated when they become the focus of attention, or are linked to the focus of attention, then fade into a state of deactivation as other traces move to the center of focus. Working memory is said to be a "matter of degree" rather than an all or none state, in that at any moment a trace might be the focus of attention (and thereby be at a peak activation level) or it might be continuously fading from attention, if, for example, it was the focus a few seconds earlier.

The application of this model has resulted in tests of working memory capacity that look quite distinct from those based on the processing workspace model. Figure 7.3 illustrates a test developed by Woltz (1988) to reflect individual differences in activation capacity. In this test subjects are presented a series of word pairs and are requested to determine whether or not the words are synonyms. Occasionally, words are repeated one, two, four, or eight items later. As Figure 7.3 shows, mean response time is 1,265 ms if neither of the words were shown before, but that time is reduced by 191 ms if one of the words was encountered on the previous item, and by 107 ms if one of the words was encountered eight items ago. The interpretation is that the word encountered even eight items ago is still more highly active than it would be as its true resting state, and therefore is processed faster. Woltz argues that individual differences in the response time facilitation effect reflect differences in activation capacity.

Given that we can define working memory capacity in two distinct ways, an important next question is, What is the empirical relationship between the two kinds of measures, and even more importantly, what is their relationship to learning? Cognitive analyses of learning tasks (Anderson, 1987;

Figure 7.3
Measuring Memory Activation Capacity

Example Items

fate	destiny
humid	damp
complain	thunder
humid	damp
polite	courteous
polite	kindle
astonish	unstable
conquer	arrange
visitor	guest
vacant	empty
complain	gripe

Measures Obtained

1. Verbal Information Processing Speed

M = 1265 ms; SD = 326 ms

2. Residual Activation Strength

Lag of Repeated Item	Mean Savings	S.D. Savings
1	191 ms	215 ms
2	124 ms	229 ms
3	108 ms	214 ms
4	107 ms	216 ms

Pirolli & Anderson, 1985) such as mathematics learning or learning a computer programming language, suggest that the limiting factor in learning is the working memory bottleneck. But the proof of this assertion is often rather theoretical, based on a rational analysis of learning task requirements, supplemented by a formal computer simulation of learning processes. An individual differences analysis of the role of working memory in learning can be a useful supplement to this kind of formal analysis, and is a fair test of the theoretical claim (Underwood, 1975). Thus we have recently begun investigating the relationship between working memory capacity as measured by tests such as those displayed in Figures 7.2 and 7.3 and performance in realistic learning contexts. We currently are investigating the acquisition of electronics troubleshooting and computer programming skills (Kyllonen, Soule, & Stephens, 1988), and other procedural learning tasks (Woltz, 1988). In all cases we find that working memory, as indicated by both the processing workspace and activation capacity measures, is a strong predictor of learning outcomes. These analyses are beginning to clarify our understanding of working memory. These studies also suggest that the particular tests of working memory capacity that we have already developed (Figures 7.2 and 7.3) are solid candidates for inclusion in future testing batteries.

Knowledge

In our four source framework for cognitive skill assessment, we refer to declarative knowledge and procedural skill as *enablers*. It has been argued that the main contribution from cognitive psychology to the new generation of psychological tests is in how we now can assess the *mediators*—information processing speed, and working memory capacity—rather than the enablers. The idea behind this thinking is that existing tests already do an adequate job at sampling the breadth of an individual's knowledge. For example, existing vocabulary tests probably are fair samples of what a person knows (although faceted vocabulary tests with a consistent sampling scheme are probably even better, Marshalek, 1981; Anderson & Freebody, 1979; Cronbach, 1942). And the Armed Services Vocational Aptitude Battery (the ASVAB) includes a number of subtests—Auto and Shop Knowledge, Mechanical Comprehension, Electrical Knowledge—that are clearly designed to sample the breadth of technical knowledge a student brings to the test.

Thus in much of our research, the measurement of knowledge has played a rather small role, especially when considered against the backdrop of its critical role in current cognitive theories generally. In experiments conducted to date, we have assessed knowledge primarily as a means for statisically controlling its effects; our main goal has been to investigate the mediator variables, which is best done by holding the knowledge effect constant.

Perhaps the reason we have failed to progress in assessing the role of knowledge in learning is that our learning tasks have purposely been rather domain-independent. It may be that advances in understanding the role of knowledge will be forthcoming only once we begin our actual complex learning experiments (described in the next section). Still, there is a considerable body of cognitive research conducted over the last ten years that enables speculations.

We propose that an individual's declarative knowledge base may be characterized along four general dimensions: *depth, breadth, accessibility* (durability), and *organization*. *Depth* refers to the amount of domain-specific conceptual knowledge possessed by the individual. Conventional achievement tests, and especially job surveys as they are employed in assessing trainee or apprentice status, are designed to tap this dimension of declaritive knowledge. *Breadth* refers to the amount of general factual knowledge available. Current intelligence tests, such as the Wechsler Adult Intelligence Test (WAIS), include an Information subtest designed to probe breadth of knowledge. And vocabulary tests can also be seen as measures of breadth of knowledge. *Accessibility* refers to the strength of the knowledge, that is, the likelihood (and the speed with which) it will be accessed in a situation in which it could be used. Accessibility is both a general characterization of all knowledge an individual possesses, and a specific parameter of every fact in the knowledge base. Accessibility is also a dynamic property of specific knowledge in that it weakens with disuse and grows stronger with practice. *Organization* refers to the relations and connections among the facts in the knowledge base. A considerable body of research in cognitive science has grown around the idea that acquiring expertise in a domain involves the reorganization of facts in the domain (for example, Lesgold, 1984).

Various methods have been developed to tap these knowledge dimensions. Clustering and scaling methods have been used to map the organization of knowledge in numerous domains such as physics (Chi, Glaser, & Rees, 1982), biology (Stephens, 1987), computer science (Adelson, 1981), psychology (Fabricious, Schwanenflugel, Kyllonen et al., 1987), and so on. Typically, a student is asked to judge the similarity of two concepts selected from the domain. Clustering and scaling methods are used to capture the underlying model used by the student to generate the similarity judgments.

There are many ways to tap accessibility of knowledge: We have used the sentence verification technique extensively (for example, Tirre, Royer, Greene, & Sinatra, 1987). Learning in the typical training situation involves listening to a lecture or reading a text, then solving problems based on the material just heard or read. The sentence verification technique is designed to probe the amount of material the learner was able to successfully encode and store in long-term memory following the listening or reading episode. The technique requires learners to discriminate paraphrases of sentences just read from modified paraphrases that are inconsistent with what was just read. Other techniques such as the cloze procedure (fill-in the blanks of sentences extracted from the preceding text) have been used for a similar purpose (Landauer, 1986). We are currently using the sentence verification technique for tracking the accumulation of declarative knowledge during the course of short (45 minutes) instructional episodes in computer programming (Kyllonen, Soule, & Stephens, 1988) and electronics troubleshooting.

Even the measurement of the depth and breadth dimensions of knowledge may benefit from recent work in cognitive science. The most innovative recent developments in probing declarative knowledge have been pursued by researchers concerned with achievement testing (Fredericksen et al., in press; Glaser, Lesgold, & Lajoie, in press; Haertel, 1985; Lesgold, Bonar, & Ivill, 1987). Glaser et al. (in press) point out that current methods, typically five-alternative multiple choice tests, suffer two key drawbacks. First, the alternatives cannot possibly accommodate all the possible misconceptions a student could possess, and thus are of limited diagnostic utility. Second, the alternatives may give away the answer, as has been shown in other realms.

Glaser et al. discuss the potential of cognitive approaches to knowledge assessment, which in contrast rely primarily on a very detailed analysis of verbal protocols extracted from students struggling with new material or applying what they have already learned. Analysis of these kinds of protocols has played a critical role in the development of a cognitive science (Ericsson & Simon, 1984), and serves as the primary basis for what Glaser, Lesgold, Lajoie et al. (1985) have dubbed *cognitive task analysis*. The problem with wholesale adoption of the technique at this time is expense: protocol analyses are costly in both subject and interviewer time, and are therefore not appropriate for inclusion in a test battery.

But Glaser et al. suggest an ingenious compromise between conventional and protocol methods. In their *hiererchical menus methodology,* students select alternatives from a series of linked menus. For example, if there are five alternatives to each menu, and there are three levels of linked menus, there can be $5^3 = 125$ response alternatives. This is superior to simply presenting 125 alternatives on screen, for two reasons. First, selecting from among 125 alternatives would impose a severe processing load on subjects, and would invite nuisance individual-difference variation in strategy selection and test-taking strategy. Second, the hierarchical arrangement can closely mirror the way in which a student is thinking about a problem, in a kind of top-down fashion.

Thus far, this approach to probing an individual's knowledge has been employed in one of the CoLoSsaL tutoring systems (BRIDGE, Bonar and Cunningham, 1986). BRIDGE, which teaches learners how to program in Pascal, presents general programming problems to be solved. At the top level (the first set of questions) the alternatives are general categories or general approaches to the problem (for example, "add something together" or "keep doing something"). Once the student selects a category, she is presented a list of alternatives that refine the category selection, and so on, until a fully specified answer is selected. From pilot testing using Air Force subjects, the method has proved general enough to accommodate the vast majority of potential responses to particular programming problems, and therefore the approach seems highly promising as a way of assessing knowledge status in the student.

To summarize, although we have not yet fully explored the domain of how to probe a learner's declarative knowledge base, we have made some important initial steps. It is likely that as we begin further testing in the more complex tutoring systems environments, the methods described in this section will be refined further.

Skills

We define skills or *procedural knowledge* as it is referred to in the cognitive science literature, fairly informally as any unit of knowledge that is typically or would likely be presented in production system simulations in the form of an if-then rule or series of if-then rules. This is any knowledge or skill the student has that might bear directly on problem solving ("how-to knowledge"). Procedural skill varies widely along the generality dimension: at the most general level are problem-solving heuristics or approaches, such as working backwards, means-end analysis, or persisting in the face of uncertainty. At the opposite end of the continuum are very specific procedures, such as moving the cursor to position 12, 45 when required to delete a character at position 12, 45.

One fairly consistent finding in cognitive research is that while specific procedures are trainable, general procedures are quite resistant to modification. This finding is certainly not due to a shortage of attempts to modify general skills. Kulik, Bangert-Downs, and Kulik (1984) reviewed over 50 studies of the effects of extensive coaching for the Scholastic Aptitude Test (SAT). They concluded that the effects, for even long-term training, were quite small (approximately one-sixth to one-third of a standard deviation, or 17 to 34 points). Venezuela's Project Intelligence (Herrnstein, Nickerson, de Sanchez, Swets, 1986) may be seen similarly as somewhat disappointing. Despite an ambitious project in which domain-free thinking skills were taught four days a week, in 45-minute lessons, for an entire year, the actual changes experienced on standard measures of cognitive skill (intelligence tests) were quite minuscule (about .3 sd). These findings should not have come as any great surprise. Attempts to have students transfer general problem-solving approaches to superficially distinct but isomorphically identical problems have repeatedly failed (for example, Brown & Campione, 1978; Simon & Hayes, 1976).

On the other hand, there is good evidence for the modifiability of specific skills, especially in context. Schoenfeld (1979) has shown how training in mathematical heuristics (for example, draw a diagram, simplify the problem, test the limiting case) can facilitate subsequent problem solving so long as the instruction is wedded tightly to the domain material simultaneously being taught. Recent analyses of transfer of training have shown that skill transfer is excellent and quite predictable when the skills transferred are close at some conceptual level to the new skills (Anderson, 1987; Kieras & Bovair, 1986).

The implications for these two results on testing are apparent. On the one hand, specific procedural knowledge is rather easily modifiable and therefore ought to perhaps be trained rather than tested for, at least in the personnel selection and classification context. Recent work on diagnostic monitoring (Frederiksen et al., in press; Lesgold et al., 1987) shows how tests can be used to tailor instruction and is thus appropriate for this purpose. On the other hand, the availability of general procedural knowledge should have an important predictive relationship to learning ability, and it seems to be fairly immutable. General procedural knowledge therefore is an ideal capability to test for in entrance (selection and classification) testing. It is interesting that researches from very diverse perspectives—psychometric (Cattell, 1971), information processing (Sternberg, 1981a), and artificial intelligence (Schank, 1980)—have argued consistently for the importance of the ability to cope with novel problems as a key aspect of intelligence, and therefore as an ideal candidate for inclusion in aptitude test batteries.

Do we now test for general procedural knowledge, or general problem-solving skills? As was the case with declarative knowledge, there certainly are in existence paper and pencil tests that would appear to tap very general problem-solving skills, Raven's Progressive Matrixes being an excellent example. And about seven years ago, Educational Testing Service began supplementing its existing Verbal and Quantitative portions of the Graduate Record Examination with a new test of Analytic ability (Wilson, 1976). However, the closest the ASVAB comes to testing general problem-solving ability is with the Arithmetic Reasoning subtest. This test consists of story problems such as "How many 36-passenger buses will it take to carry 144 people?" (DoD, 1984). Recall that the Arithmetic Reasoning test loaded highly on the working memory factor in the Christal (1987) study, which suggests an intriguing research question: Just what is the relationship between working memory and procedural skill?

We can think of working capacity as mediating the development and efficiency of general problem-solving strategies. But an alternative view of the relationship between the two constructs assigns the central role to working memory. Baddeley (1987) has proposed a model of working memory consisting of various slave storage subsystems (for storing linguistic information, spatial information, etc.), along with a central executive, which monitors and coordinates the activities of the subsidiary storage systems. Executive skill, then, is skill in monitoring one's problem-solving processes, adapting to changing task requirements, successfully executing general problem-solving strategies, allocating resources where they are needed, and more generally, changing processing strategy in accordance with changes in processing demands.

In this way the executive can be seen as the most important component of working memory. Yet, while we have a reasonable understanding of how the subsidiary storage systems function, according to Baddeley, the workings of the central executive still remain largely a mystery. An important and exciting research direction is to begin devising means for measuring executive skill and thereby begin unpacking that mystery.

Modeling Learning Skills

Learning Skills Taxonomy

If we can adequately measure knowledge and the various skills associated with the four sources, an important next step in the research program is to demonstrate the relationship between those scores generated from a trainee's interaction with a learning task. We believe that learning should be expressible in terms of (that is, predictable from) the underlying components, but it is necessary to prove that this is the case.

Much of our research until fairly recently has used grossly simplified learning tasks as criterion measures against which to validate the new cognitive abilities measures. For example, in the Kyllonen-Tirre-Christal study, performance on various paired-associates tests was used as criteria, and in other studies, we have employed comparably simple, short-term learning tasks. The logic underlying this decision is twofold. First, we are concerned with developing rigorous models of the aptitude-learning-outcome relationship, and simple, short-term learning tasks afford more control over the instructional environment. But second, we believe that the kinds of learning involved in even these simple tasks is at some fundamental level the same as that involved in more realistic learning situations. Or, conversely, even apparently complex classroom learning can be analyzed and decomposed into a series of much simpler learning acts.

If we accept the notion that even complex learning tasks can be broken down into their constituent learning activities, then it obviously would be useful to specify the nature of those basic learning activities. One proposal that has been useful in our work, based largely on Anderson's (1987) three-stage model of skill acquisition, is represented on the right side of Figure 7.1. The idea is that cognitive skills develop through an initial engagement of declarative learning processes ("memorizing the steps"), followed by an engagement of proceduralization processes ("executing the steps"), then finally refinement processes ("automatizing the steps"). As Figure 7.4 shows, different performance measures will be sensitive to the course of skill development at various points along the way. When first learning a skill, many mistakes will be made, and accuracy measures will be the most sensitive indicators of skill development. Later, when the skill is known, few mistakes will be made, and performance time measures will be the most

Figure 7.4
Knowledge/Skill Acquisition

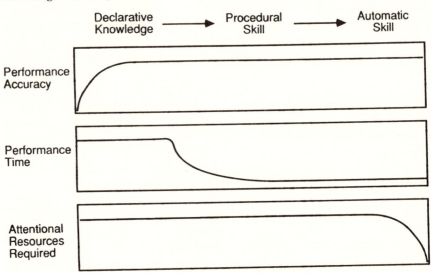

sensitive indicators. Still later, performance time will approach a minimum as the target skill becomes increasingly automatized, but there might still be considerable variability in whether (and how much) other processing can be occuring while the target skill is being executed.

We (Kyllonen & Shute, in press) recently elaborated on this simple taxonomy in proposing that in addition to the status of the skill (that is, whether the skill is in a declarative, procedural, or automatic state, which we identified as the *knowledge-type* dimension), learning could be classified along three other dimensions: the *learning environment,* the *domain,* and the learner's *cognitive style.*

The *learning environment* specifies the nature of the inference process required by the student: The simplest learning act involves rote memorization. Learning by actively encoding, by deduction, by analogically reasoning, by refinement through reflection following practice, by induction from examples, and by observation and discovery, involve successively more complex processing on the part of the learner. The second dimension, the resulting *knowledge-type,* as indicated above, specifies whether the product of the learning act is a new chunk of declarative knowledge (a new fact, or body of facts) or new procedural knowledge (a rule, a skill, or a mental model). The third dimension, the *domain,* refers to whether learning is occuring in a technical, quantitative domain, or a more verbal, nontechnical domain. Together, these three dimensions specify a particular kind of learning act. The fourth dimension, the learner's *cognitive style,* is a property of the learner, rather than the instructional situation per se. But we included it in recognition of the possibility that we cannot be certain on any task of what learning skill is being assessed unless we consider how the learner is approaching the task.

Our proposal. which has not in any sense been put to the test, is that the taxonomy should be useful in two ways. First, it provides a sampling space from which we may draw learning tasks. The goal of the LAMP effort is to model learning ability using cognitive skill measures: The taxonomy specifies the range of learning tasks for which we must develop adequate models. Second, in reverse fashion, the taxonomy specifies the kinds of micro-level learning acts that combine to make complex learning. This is the aspect of providing a task analysis tool. Our idea is that we can inspect the requirements of any complex learning situation, in the classroom or in front of a computer, and specify what learning acts are occurring. Given any instructional exchange, we can find a cell in the taxonomy that represents that exchange.

Complex Learning Skills Laboratory (CoLoSsaL)

One potential stumbling block for any program like ours is that it is not easy to monitor progress. To determine whether our new, innovative measurement methods are valid predictors of learning success it is necessary to observe students engaged in learning. Two appraoches have traditionally been taken. One is to validate the new tests against some criterion reflecting success in operational training, such as final course grade-point average. The benefit of this approach is that inferences from the research are direct, but there are a number of drawbacks: data collection is extremely slow, instructor quality is highly variable and may interact with learner characteristics in affecting learning outcomes, and there is no allowance for manipulating the learning task in any way so as to allow "what if" questions regarding validity (for example, "what if the instructor encouraged more questions, would that differentially affect student outcomes?").

The second approach is to simplify the learning task so that it is under the experimenter's control and can be administered within a single session. With complete control over the learning task, one can ask and test "what if" questions easily. Unfortunately, in so modifying the learning task the researcher cannot necessarily continue to assume that the instruments shown to be valid in the experimental context will prove to be valid in predicting success in more realistic learning situations.

Our solution to the validity problem represents a compromise between these two positions. We are currently designing intelligent computerized tutoring systems to teach computer programming, electronics troubleshooting, and flight engineering in 56-hour mini-courses (LRDC, 1987). In addition, we will add new mini-courses over the next several years. The tutoring systems are being designed to produce a rich variety of indexes of the learner's curriculum knowledge and his or her progress in acquiring the new knowledge and skills being taught. The tutoring systems are sufficiently flexible so that it is easy to modify the instructional strategy and thus ask "what if" questions. The learning involved on the other hand is not trivial. It has been estimated that one hour of tutored instruction is equivalent to approximately four hours of regular classroom instruction (Anderson, Boyle, & Reiser, 1984), and thus these mini-courses are quite extensive. A major goal of our current research efforts is to use the taxonomy to generate the most expressive indexes of the student's learning experience.

We envision a broad range of research questions that can be addressed once we begin gathering data with these kinds of learning indexes. First, the indexes can serve as alternatives to end-of-course achievement test scores as criteria for validating new cognitive aptitude tests. An index such as "probability of remembering an instructional proposition (as a function of the amount of study and presentation lag)" is more precise and potentially more general than a broad achievement test score. Such a fine breakdown of the learning experience also permits enhanced analyses among the indexes themselves. For example, we can begin investigating more precisely questions concerning the relationship between initial knowledge acquisition and the subsequent ability to turn that knowledge into problem-solving skill, or the ability to tune that skill with more problem-solving experience.

Finally, developing rich profiles of an individual learner's strengths and weaknesses in the form of elaborate assemblies of learning indexes should permit a reassessment of the aptitude-treatment-interaction (ATI) idea (Cronbach & Snow, 1977). Probably, the inconclusiveness of past ATI research can be traced to the employment of global aptitude indexes and global learning outcome measures along with pragmatic limitations on instructional variation. The tutoring systems being developed overcome these limitations by generating richer traces of a learner's path through a curriculum, and by being sufficiently flexible so as to allow potentially unlimited variations in how instruction is presented.

SUMMARY AND CONCLUSIONS

This report has outlined some of the research activities underway as part of the Air Force's Learning Abilities Measurement Project (LAMP). The major goal of the project is to devise new models of the nature and organization of human abilities with the long-term goal of applying those models to improve current personnel selection and classification systems.

As an approach to this ambitious undertaking, we have divided the activities of the project into two categories. The first category is concerned with identifying fundamental learning abilities by determining how learners differ in their ability to think, remember, solve problems, and acquire knowledge and skills. From research already completed, we have established a four-source framework that assumes that observed learner differences are due to differences in information processing efficiency, working memory capacity, and the breadth, extent, and accessibility of conceptual knowledge and procedural and strategic skills.

The second category of research activities is concerned with validating new models of learning abilities. To do this we are building a number of computerized intelligent tutoring systems that serve as mini-courses in technical areas such as computer programming and electronics troubleshooting. A major objective of this part of the program is to develop principles for producing indicators of student learning progress and achievement. These indicators will serve as the learning outcome measures against which newly developed learning abilities tests will be evaluated in future validation studies. The indicators also will be applied in studies that investigate the dynamics of knowledge and skill acquisition and in studies that attempt to optimize instruction so as to capitalize on and compensate for learner strengths and weaknesses.

REFERENCES

Adelson, B. (1981). Problem solving and the development of abstract categories in programming languages. *Memory & Cognition, 9.* 422-433.

Anderson, J. R. (1987). Skill acquisition: Compilation of weak-method problem solutions. *Psychological Review, 94,* 192-210.

Anderson, J. R., Boyle, C. F., and Reiser, B. J. (1984). Intelligent tutoring systems. *Science, 228,* 456-462.

Anderson, J. R., Farrell, R., and Sauers, R. (1984). Learning to program in LISP. *Cognitive Science, 8,* 87-129.

Anderson, J. R., and Jeffries, R. (1985). Novice LISP errors: Undetected losses of information from working memory. *Human-Computer Interaction, 22,* 403-423.

Anderson, R. C., and Freebody, P. (1979). *Vocabulary knowledge* (Report No. 136). Champaign: University of Illinois, Center for the Study of Reading.

Baddeley, A. D. (1968). A 3 min reasoning test based on grammatical transformation. *Psychonomic Science, 10,* 341-342.

———. (1987). *Working memory.* New York: Academic Press.

Baddeley, A. D., and Hitch, G. (1974). Working memory. In G. Bower (Ed.), *Advances in learning and motivation, Vol. 8.* New York: Academic Press.

Bonar, J. (1984). *Understanding the bugs of novice programmers.* (Doctoral dissertation, University of Massachusetts.)

Bonar, J. G., and Cunningham, R. (1986). *BRIDGE: An intelligent tutor for thinking about programming* (Report No. 1). Pittsburgh, PA: University of Pittsburgh, Learning Research and Development Center.

Brown, A. L., and Campione, J. C. (1978). Memory strategies: Training children to study strategically. In H. Pick, H. Lebowitz, J. Singer, A. Steinschneider, and H. Stevenson (Eds.), *Application of basic research in psychology.* New York: Plenum.

Brown, J. S., and Burton, R. R. (1978). Diagnostic models for procedural bugs in basic mathematical skills. *Cognitive Science, 2,* 155-192.

Carrol, J. B. (April, 1981). *Individual difference relations in psychometric and experimental cognitive tasks* (Report No. 163). Chapel Hill: University of North Carolina.

Cattell, R. B. (1971). *Abilities: Their structure, growth, and action.* Boston: Houghton Mifflin.

Chi, M. T. H., Glaser, R., & Rees, E. (1982). Expertise in problem solving. In R. J. Sternberg (Ed.), *Advances in the psychology of human intelligence.* Hillsdale, NJ: Erlbaum.

Christal, R. E. (1981). *The need for laboratory research to improve the state-of-the-art in ability testing.* Paper presented at the National Security Industrial Association-DoD Conference on Personnel and Training Factors in Systems Effectiveness, San Diego, CA.

_____ . (1987). *A factor-analytic study of tests of working memory.* Unpublished manuscript.

Collins, A. (1986). High pay-off research areas in the cognitive sciences. In T. G. Sticht, F. R. Chang, and S. Wood (Eds.), *Advances in Reading/ Language Research: Cognitive Science and Human Resources Management.* Greenwich, CN: JAI Press.

Cronbach, L. J. (1942). An analysis of techniques for diagnostic vocabulary testing. *Journal of Educational Research, 36,* 206–217.

Cronbach L. J., and Snow, R. E. (1977). *Aptitudes and instructional methods: A handbook for research on interactions.* New York: Irvington.

Daneman, M. and Carpenter, P. A. (1980). Individual differences in working memory and reading. *Journal of Verbal Learning and Verbal Behavior, 19,* 450–466.

Department of Defense. (1984). *Test manual for the Armed Services Vocational Aptitude Battery* (DoD, 1340.12AA). North Chicago, IL: U. S. Military Entrance Processing Command.

Detterman, D. K. (1980). Does "g" exist? *Intelligence, 6,* 99–108.

_____ . (1986, November). *Basic cognitive processes predict IQ.* Paper presented at the Twenty-seventh Annual Meeting of the Psychonomic Society, New Orleans, LA.

Donaldson, G. (1983). Confirmatory factor analysis models of information processing stages: An alternative to difference scores, *Psychological Bulletin, 94,* 143–151.

Embretson, S. (in press). Diagnostic testing by measuring learning processes: Psychometric considerations for dynamic testing. In N. Fredericksen, A. Lesgold, R. Glaser, and M. Shafto (Eds.), *Diagnostic monitoring of skill and knowledge acquisition.* Hillsdale, NJ: Erlbaum.

Ericsson, K. A., and Simon, H. A. (1984). *Protocal analysis: Verbal reports as data.* Cambridge, MA: MIT Press.

Fabricious, W. F., Schwanenflugel, P. J., Kullonen, P. C., Barclay, C., and Denton, M. (1987, April). *Developing concepts of the mind: Children's and Adults' Representations of mental activity.* Paper presented at the meeting of the Society for Research in Child Development, Baltimore, MD.

Fairbank, B. B., Jr. (1988). *Mathematical analysis of reaction-time distribution.* Paper presented at the Human Factor Society Annual meeting, Anaheim, CA.

Fredericksen, J. R., Weaver, P. A., Warren, B. M., Gillotte, H. P., Rosebery, A. S., Freeman, B., and Goodman, L. (1983, March). *A componential approach to training reading skills: Final Report* (Report No. 5295). Cambridge, MA: Bolt, Baranek, & Newman.

Fredericksen, N., Lesgold, A., Glaser, R., and Shafto, M. (Eds.). (in press). *Diagnostic monitoring of skill and knowledge acquisition.* Hillsdale, NJ: Erlbaum.

Gitomer, D. (1984). *A cognitive analysis of a complex troubleshooting task.* Unpublished doctoral dissertation, University of Pittsburgh.

Glaser, R. (1985, October). *The integration of instruction and testing.* Paper presented at the ETS Invitational Conference. Princeton, NJ: Educational Testing Service.

Glaser, R., Lesgold, A., and Lajoie, S. (in press). Toward a cognitive theory for the measurement of achievement. In R. R. Ronning, J. Glover, J. C. Conoley, and J. C. Witts (Eds.), *The influence of cognitive psychology on testing, Buros/Nebraska Symposium on Testing, Vol. 3.* Hillsdale, NJ: Erlbaum.

Glaser, R., Lesgold, A. M., Lajoie, S., Eastman, R., Greenberg, L., Logan, D., Magone, M., Weiner, A., Wolf, R., and Yengo, L. (1985, October). *Cognitive task analysis to enhance technical skills training and assessment* (Tech. Report). Pittsburgh, PA: University of Pittsburgh, Learning Research and Developmental Center.

Haertel, E. (1985). Construct validity and criterion-referenced testing. *Review of Educational Research, 55,* 23–46.

Herrnstein, R. J., Nickerson, R. S., de Sanchez, M. and Swets, J. A. (1986). Teaching thinking skills. *American Psychologist, 41,* 1279–1289.

Hitch, G. J. (1978). The role of short-term working memory in mental arithmetic. *Cognitive Psychology, 10,* 302–323.

Hockley, W. E. (1984). Analysis of response time distributions in the study of cognitive processes. *Journal of Experimental Psychology: Learning, Memory, and Cognition, 10,* 598–615.

Hunt, E. (1982). Toward new ways of assessing intelligence. *Intelligence, 6,* 231–240.

Hunt, E. B., Frost, N., and Lunneborg, C. (1973). Individual differences in cognition: A new approach to intelligence. In G. Bower (Ed.), *The psychology of learning and motivation: Advances in research and theory (Vol. 7).* New York: Academic Press.

Hunt, E. B., and Pellegrino, J. W. (1984). *Using interactive computing to expand intelligence testing: A critique and prospectus* (Report No. 84–2). Seattle: University of Washington, Department of Psychology.

Jensen, A. R. (1982). Reaction time and psychometric g. In H. J. Eysenck (Ed.), *A model for intelligence.* New York: Springer-Verlag.

Kieras, D. E., and Bovair, S. (1986). The acquisition of procedures from text: A production system analysis of transfer of training. *Journal of Memory and Language, 25,* 507–524.

Kulik, J. A., Bangert-Downs, R. L., and Kulik, C. L. C. (1984). Effectiveness of coaching for aptitude tests. *Psychological Bulletin, 95,* 179–188.

Kyllonen, P. C. (1985). *Dimensions of information processing speed* (AFHRL-TP-84-56, AD-A154-778). Brooks AFB, TX: Manpower and Personnel Division, Air Force Human Resources Laboratory.

_____ . (1986). Theory-based cognitive assessment. In J. Zeidner (Ed.), *Human productivity enhancement: Organizations, personnel, and decision making, Vol. 1* (pp. 338–381). New York: Praeger.

_____ . (1987). *Componential analysis of semantic matching.* Unpublished manuscript, University of Georgia, Athens.

Kyllonen, P. C., and Shute, V. J. (in press). Learning indicators from a taxonomy of learning skills. In P. Ackerman, R. J. Sternberg, and R. Glaser (Eds.), *Learning and individual differences.* New York: Freeman.

Kyllonen, P. C., Soule, C., and Stephens, D. (1988).*The role of working memory and general problem solving skill in acquiring computer programming skill,* Unpublished manuscript.

Kyllonen, P. C., and Stephens, D. (1988). *The role of working memory and accretive learning processes in learning logic gates,* unpublished manuscript.

Kyllonen, P. C., Tirre, W. C., and Christal, R. E. (1984, November). *Processing determinants of associative learning.* Paper presented at Psychonomic Society, San Antonio, TX.

_____ . (1985). *The speed-level problem reconsidered.* Manuscript submitted for publication.

_____ . (1987). *Knowledge and processing speed as determinants of associative learning.* Manuscript submitted for publication.

Landauer, T. K. (1986). How much do people remember? Some estimates of the quantity of learned information in long-term memory. *Cognitive Science, 4,* 477–494.

Learning Research and Development Center. (1987). *Research in Intelligent CAI at the Learning Research and Development Center of the University of Pittsburgh.* Pittsburgh, PA: University of Pittsburgh, LRDC.

Lesgold, A. (1984). Acquiring expertise. In J. R. Anderson and S. M. Kosslyn (Eds), *Tutorials in learning and memory: Essays in honor of Gordon Bower* (pp. 31-64). San Francisco: Freeman.

Lesgold, A., Bonar, J., and Ivill, J. (1987, March). *Toward intelligent systems for testing* (Report No. LSP-1). Pittsburgh, PA: University of Pittsburgh, LRDC.

Marshalek, B. (1981, May). *Trait and process aspects of vocabulary knowledge and verbal ability* (Report No. 15). Stanford University, School of Education, Aptitude Research Project.

McClelland, J. L. (1979). On the time relations of mental processes: An examination of systems of processes in cascade. *Psychological Review, 86,* 287–330.

Naveh-Benjamin, M., McKeachie, W. J., Lin, Y. G., and Tucker, D. G. (1986). Inferring students' cognitive structures and their development using the "Ordered tree technique." *Journal of Educational Psychology, 78,* 130–140.

Neisser, U. (1967). *Cognitive psychology.* New York: Appleton-Century-Crofts.

Pellegrino, J. W., and Glaser, R. (1979). Cognitive correlates and components in the analysis of individual differences. In R. J. Sternberg and D. K. Detterman (Eds.), *Human intelligence: Perspectives on its theory and measurement* (pp. 61–88). Norwood, NJ: Ablex.

Pirolli, P. and Anderson, J. R. (1985). The role of learning from examples in the acquisition of recursive programming skill. *Canadian Journal of Psychology, 39,* 240–272.

Posner, M. I., and McLeod, P. (1982). Information processing models—In search of elementary operations. *Annual Review of Psychology, 33,* 477–514.

Ratcliff, R. and Murdock, B. B., Jr. (1976). Retrieval processes in recognition memory. *Psychological Review, 83,* 190–214.

Rose, A. M., and Fernandez, K. (1977). *An information processing approach to performance assessment, I. Experimental investigation of an information processing performance battery* (Report No. 1). Washington, DC: American Institute for Research.

Schank, R. C. (1980). How much intelligence is there in artificial intelligence? *Intelligence, 4,* 1–14.

Schoenfeld, A. H. (1979). Explicit heuristic training as a variable in problem solving performance. *Journal for Research in Mathematics Education, 10,* 173–187.

Schvaneveldt, R. W., Durso, F. T., Goldsmith, T. E., Breen, T. J., Cooke, N. M., Tucker, R. G., and DeMaio, J. C. (1985). Measuring the structure of expertise. *International Journal of Man-Machine Studies, 23,* 699–728.

Simon, H. A., and Hayes, J. R. (1976). The understanding process: Problem isomorphs. *Cognitive Psychology, 8,* 165–190.

Snow, R. E. (1979). Theory and method for research on aptitude processes. In R. J. Sternberg and D. K. Detterman (Eds.), *Human intelligence: Perspectives on its theory and measurement* (pp. 105–137). Norwood, NJ: Ablex.

Stephens, D. L. (1987). *Use of cognitive structure in predicting test achievement and ideational creativity in biology students.* Unpublished master's thesis, Department of Educational Psychology, University of Georgia, Athens, GA.

Sternberg, R. J. (1977). *Intelligence, information processing, and analogical reasoning: The componential analysis of human abilities.* Hillsdale, NJ: Erlbaum.

———. (1980). Nothing fails like success: The search for an intelligent paradigm for studying intelligence. *Journal of Educational Psychology, 73,* 142–155.

———. (1981a). Intelligence and nonentrenchment. *Journal of Educational Psychology, 73,* 1–16.

———. (1981b). Testing and cognitive psychology. *American Psychologist, 36,* 1181–1189.

Sternberg, R. J., and Powell, J. (1983). Comprehending verbal comprehension. *American Psychologist, 38,* 878–893.

Tirre, W. C., Royer, J. M., Greene, B. A., and Sinatra, G. M. (1987, April). *Assessing on-line comprehension in a computer based instruction environment.* Paper presented at the meeting of the American Educational Research Association, Washington, D. C.

Underwood, B. J. (1975). Individual differences as a crucible in theory construction. *American Psychologist, 30,* 128–134.

Wilson, K. E. (1976). *The GRE Technical Manual.* Princeton, NJ: Educational Testing Service.

Woltz, D. J. (1988). An investigation of the role of working memory in procedural skill acquisition. *Journal of Experimental Psychology: General, 117,* 319–331.

Woltz, D. J., and Christal, R. E. (1985, April). *Working memory.* Paper presented at the Western Psychological Association Annual Convention, San Jose, CA.

8 Computer-Controlled Assessment of Static and Dynamic Spatial Reasoning

James W. Pellegrino
Earl B. Hunt

OVERVIEW

Over the last ten years, we and others have been involved in research which focuses on information processing, componential analyses of individual differences in cognitive abilities. As a result, there are now sophisticated theories and models of a wide range of cognitive performances with many interesting results about the nature of developmental and individual differences in various aspects of cognition. The domains of inquiry have gradually expanded to include componential analyses of virtually all the major aptitude factors identified in psychometric research and typically assessed by general and specific aptitude tests (for example, Sternberg, 1985a, 1985b). An example is spatial ability which represents the focus of this chapter (for example, Lohman, Pellegrino, Alderton, & Regian, 1986; Pellegrino, Alderton, & Shute, 1984; Pellegrino & Kail, 1982). Our goal is not to review what has been learned from various individual studies pursuing information processing, componential analyses of spatial aptitude. Rather we are concerned with issues in applying such knowledge to practical problems of assessment.

For some time, we and others have argued that componential analysis approaches to the study of individual differences have much to offer in the area of assessment. We have tried to demonstrate in numerous studies that this approach can be used to understand the sources of individual differences in cognitive abilities. In this regard we have been relatively successful. However, we have failed to "put our money where our collective mouths were." In virtually every study using a componential analysis approach there has been no real concern about the practicality of implementing this assessment approach on any large-scale basis. The method has been applied for the purpose of addressing scientific questions, where it has been successful, but there is limited evidence of its practicality for large-scale testing.

We will show that for the domain of spatial ability, componential analysis procedures are feasible when conducted in the context of computer-based testing. Furthermore, computer-based testing alows for assessment of dynamic spatial reasoning skills. The latter have heretofore never been assessed. In the next section we briefly review some of the key issues associated with traditional and contemporary assessment of spatial abilities.

ISSUES IN THE ASSESSMENT OF SPATIAL ABILITY

Spatial ability is the ability to reason about visual scenes. An example would be the reasoning a pedestrian does when deciding to cross a busy street. The positions of moving vehicles must be moved foward in time. Another, less serious, example is the reasoning one does in deciding to move a piece into an open position in a jigsaw puzzle. Spatial ability is a basic dimension of human intellegence, clearly separate from verbal intelligence or general reasoning ability. Virtually every comprehensive theory of intelligence makes reference to the domain of spatial ability (Carroll, 1982).

Spatial ability is better thought of as a domain of abllities than as a single ability or skill. Multivariate studies of the domain have identified three major factors (Lohman, 1979; McGee, 1979). The most clearly defined factor is *spatial relations* ability which refers to the capacity to move objects "in the mind's eye," as is required when one "mentally rotates" an object about its center (Shepard & Cooper, 1982). Conventional psychometric tests of spatial relations ability include the Primary Mental Abilities Space test (Thurstone & Thurstone, 1949). A second factor is *spatial visualization*, which is best thought of as the ability to deal with complex visual problems which require imagining the relative movements of internal parts of a visual image. The jigsaw puzzle example given earlier is a good illustration. Psychometric tests that tap spatial visualization include the paper-folding (surface development) task in the Differential Aptitude Battery (DAT) (Bennett, Seashore, & Wesman, 1974) and the Minnesota Paper Form Board Test (Likert & Quasha, 1971). The third primary spatial factor is *spatial orientation* which is the ability to imagine how a stimulus or stimulus array would appear from a different perspective. Distinguishing this factor from spatial relations is often difficult.

Although three dimensions of spatial ability can be identified, they are typically correlated across individuals. Therefore, in terms of the technical aspects of multidimensional analysis, the scores from a variety of spatial ability tests may often be placed in a two- rather than a three-dimensional space. More precisely, depending upon the exact composition of the test battery, it will often be the case that three dimensions are required for an excellent fit, but that two dimensions will be "almost" sufficient.

On logical grounds alone, one might expect tests of spatial ability to predict performance in certain nonacademic fields. Consider the examples given earlier, crossing a street and constructing a jigsaw puzzle. The first problem involves estimating the time and point of arrival of objects in the visual field, a task that appears to be a primary component of many machinery-operated tasks. This component also appears to draw on spatial relations ability, as defined by typical psychometric tests. Consistent with this observation, spatial orientation tests are used to screen candidates for flight training, where one would expect spatial orientation skills to be in high demand. Spatial relations and spatial visualization tests have also been related to performance in architecture and engineering courses, where they are reliable predictors of achievement (McGee, 1979). More detailed studies have shown that spatial ability test scores can be related directly to performance on problems involving analysis of engineering drawings (Pellegrino, Mumaw, & Shute, 1985).

Outside of aviation, spatial ability tests have a rather mixed record as predictors of performance in military occupations. In spite of there being a strong logical case that the tests should predict performance in occupations involving either machinery operation or the analysis of drawings, only low to moderate correlations have been found. For that reason, spatial tests are not included in the current version of the Armed Services Vocational Aptitude Battery (ASVAB).

Elsewhere we have noted that the generally disappointing performance of spatial ability tests as predictors may be due to a technological restriction on the range of spatial abilities that can be tested (Hunt & Pellegrino, 1985). Current spatial tests virtually all depend on the conventional paper and pencil test format. This resricts the form of the test severely since the visual scenes that the examinee must reason about cannot contain moving elements. Also, while it is possible to determine how many items an examinee can pass in a fixed time, it is not possible to examine the time that a person spends on an individual item, or the time spent on various identifiable subparts of the spatial problem posed by a given test item. This is an issue because speed and accuracy in solving parts of a problem may reflect different psychological skills (Pellegrino & Kail, 1982). More generally, different people may trade-off between speed and accuracy of performance in different ways and measures of both speed and accuracy may be needed to adequately assess skill (Pachella, 1974).

We have pointed out that both of these problems may be remedied by computer-administered testing (Hunt & Pelligrino, 1985). Visual displays with moving elements (*dynamic* displays) can be presented in computer-controlled testing. In addition, computer-controlled testing makes it possible to record accuracy and latency measures each time an examinee attacks a problem. The latter advantage applies both to tasks using dynamic displays and those using displays without moving elements (*static* displays). There are, however, two concerns. Altough one can conceptualize a difference in reasoning about static versus dynamic visual displays, there is no psychometric evidence to show that this is the case. Furthermore, while it is possible to design static display problems that appear to be related to (and that are correlated with) performance on paper and pencil tests of spatial ability, it is also possible that computer-controlled testing may tap a new dimension of ability. The evidence for or against this proposition is both sparse and somewhat contradictory (Hunt & Pellegrino, 1985).

We have argued that the advent of computer-controlled testing offers a *possibility* of substantial improvements in spatial ability testing (Hunt & Pellegrino, 1985). Whether or not that possibility should be realized depends on the answers to three questions. Does computerized testing involving static displays evaluate the same abilities as conventional paper and pencil tests? Can tests using dynamic displays be designed to test a dimension of ability that is different from the abilities evaluated using static displays? Finally, do the finer measures available through computerized testing make possible more precise measurement of ability and thus better prediction of on-the-job performance? Of course, the latter is the question of most interest in applied psychology. An attempt to answer it directly, however, could be both fruitless and extremely expensive unless the first two questions are examined beforehand.

In the remainder of this chapter we describe a project, and its results which was designed to address the aforementioned issues. As part of the project we developed a computer-administered test battery containing 11 spatial reasoning tasks. Five of the tasks involved static visual displays and six involved dynamic visual displays. These computer-based tasks were administered to a large sample of young adults (N = 170) who also took a battery of conventional paper and pencil tests. One primary goal of this project was to determine whether the computer-administered static display tasks evaluated the same abilities as standard paper and pencil tests of spatial ability. The second primary goal was to determine whether tasks requiring reasoning about dynamically displayed spatial information would reveal one or more new dimensions of spatial ability, dimensions not assessed by either the paper and pencil tests or the computer-administered static display tasks. A subsidiary goal related to pursuit of the two primary goals was examination of the reliability and utility of measures of within-problem process execution speed, measures that can be obtained in computerized testing but that cannot be obtained in paper and pencil tests.

In the next two sections we briefly describe the tasks included in the computer-administered battery of spatial tests. This includes the logic behind selection and implementation of each task and the types of performance measures provided by each task. More complete details about individual task design can be found in the report by Hunt et al. (1986).

INFORMATION PROCESSING TASKS
INVOLVING STATIC DISPLAYS

The five static spatial reasoning tasks that we selected for computer implementation and administration had to fit within a set of criteria. First, we wanted tasks that had well-defined information processing characteristics, tasks that had been previously analyzed and validated with respect to underlying component processes. Second, whenever possible, the task and its derived performance/processing measures should have some history of use in the study of individual differences in spatial ability or imagery ability. Third, the full set of static tasks should represent major visual-spatial factors such as perceptual speed, spatial relations, and spatial visualization.

Perceptual Comparison

This task was selected to provide measures of perceptual speed, that is, the ability to rapidly encode figures and make visual comparisons. The computer task is based on visual comparison research conducted by Cooper (1976). On each trial a pair of random shapes is presented. The examinee is to indicate whether or not the shapes are exactly identical. The shapes are generated by connecting 6 to 14 randomly chosen points on a plane. Mismatching comparison figures are generated from a standard figure by slightly moving one or more of the original points. Two aspects of this task are of interest. First, on trials where there is a mismatch, the degree of mismatch is varied and decision time should be a monotonically decreasing function of this variable. The slope of the difference detection function provides an index of the speed (efficiency) of detecting feature differences. Second, stimulus complexity is manipulated by varying the number of points (6 to 14) used to generate the original shape. The slope of the stimulus complexity reaction time function provides an index of encoding and comparison efficiency. Measures derived from this task included average response time and accuracy, measures similar to those obtained from standard perceptual speed tests, as well as slope and intercept measures reflecting efficiency of difference detection and efficiency of encoding and comparison processes.

Mental Rotation

This task was selected to provide a measure of spatial relations ability. We used a computer-controlled mental rotation that closely resembled tasks used previously to assess individual differences in spatial relations ability (Mumaw, Pellegrino, Kail, & Carter, 1984). On each trial a pair of polygons was presented on the computer display. The task was to judge their identity and respond by pressing one of two keys. The individual trials represented the combination of two variables: angular disparity of the stimuli (ranging from 0-180 degrees) and match type (same or different). Nonmatching pairs were created by a mirror-image reversal of the stimulus. Measures of separate processes were estimated from the slopes and intercepts of the linear functions relating response time to angular disparity. General and specific latency and accuracy parameters were estimated separately for positive and negative match conditions since other research has shown that performance in the positive and negative conditions often differs (Pellegrino & Kail, 1982).

Surface Development

This task was selected to provide one example of a spatial visualization task. Spatial visualization ability is often assessed by surface development tasks which contain representations of flat, unfolded objects and completed three-dimensional shapes. The task that we used is based on previous work on individual differences in spatial visualization ability (Alderton & Pellegrino, 1984). On each trial, the individual was presented two figures. The left-hand figure showed a cube "unfolded" along its edges so that it lay flat. The base of the cube was labeled, and two or three sides were marked by dots. The right-hand figure presented a two-dimensional projection of a cube "refolded" and slightly rotated so that the top, front, and right lateral surfaces were visible. The task was to determine whether the flat figure on the left could be refolded and rotated (if appropriate) to form the cube shown on the right of the screen. Different folding patterns were used, systematically varying the number of mental folds that had to be made and the number of surfaces that had to be mentally "carried along" during such folds (for example, Shepard & Feng, 1972). Performance measures included mean latency and mean accuracy as well as slopes and intercepts of the linear functions relating response time to problem complexity as determined by the number of mental folding operations.

Integration of Detail in an Image

This task was selected to represent a second prototypical spatial visualization task. Spatial visualization ability is often assessed by form board tasks that require the integration of elements to form a composite image. Examples include cases where a set of shapes must be concatenated in a certain way to form a completed shape or puzzle. Mumaw and Pellegrino (1984) and Poltrock and Brown (1984) have developed variants of this type of task that permit assessment of various image integration processes and the capacity of the visual buffer. The task that we used is similar to that developed by Poltrock and Brown (1984). On each trial, the individual was presented an array of regular shapes with various edges marked by specific letters. The number of pieces in the array varied from three to six. The individual studied the pieces and tried to determine the composite image that would be created by appropriately aligning the pieces with corresponding edge markings. The time to perform the integration should vary with the number of shapes to be integrated. Following the integration phase, the individual was shown a completed shape and decided if it represented the correct integrated image. The slope of the latency function for image integration provides an index of the efficiency of several imagery processes while the accuracy score provides an index of buffer capacity.

Adding Detail to an Image

One of the components of Kosslyn's (1980) imagery theory is the addition or deletion of detail in a mental representation (image). To do this, several subprocesses are required. Examples are the "Put" operation, in which a component part is placed at a point in an image, and the "Find" operation, in which an image is examined to determine whether or not it contains a feature. Poltrock and Brown (1984) developed a task that provides indexes of the efficiency of these and other processes. Individuals are asked to image a base form and then add details (dots) to it at specified locations. We implemented a variant of this task for computer presentation. The base form was a six-pointed star. Trials varied in terms of the total number of dots (four to seven) to be sequentially added to the base form prior to presentation of a single composite image containing the appropriate number of dots. The subject's task was to then decide if the composite image was a correct or incorrect final product. From this task we derived several latency measures including a slope and intercept of the latency function for adding successive details. We also examined changes in accuracy as a function of problem complexity which provides an estimate of a person's visual memory buffer capacity.

INFORMATION PROCESSING TASKS INVOLVING DYNAMIC DISPLAYS

Unlike the tasks that require reasoning about static spatial information, there is little in the way of experimental literature on individual differences in motion extrapolation. Therefore, the tests we used have less history of development than the ones discussed in the section dealing with static displays. We attempted to design tasks that coincided with a rational analysis of what appears to be required in this domain of performance. The "basic" dynamic visual-spatial problem a person has to solve is to predict *where* a moving object is going and *when* it will arrive at its predicted destination. This skill can be divided into three components. First, an observer might need to remember the path the object has just traveled in order to extrapolate its future path. Second, given the path the object has traveled thus far, the observer must be able to predict or extrapolate the future path. Third, to predict when an object will arrive at a destination, the observer must extrapolate its speed. Thus, we developed three tasks that assessed these three separate components of dynamic spatial reasoning.

In addition to making judgements about the path and speed of a single moving object it would appear that many dynamic spatial problems require judgments of relative rather than absolute speed. For instance, if two objects are moving toward destinations, an observer may wish to know if they will arive at the same time or, if not, which will arrive first. We developed two tasks that assessed the ability to make such comparitive speed judgments.

While it is reasonable to break up judgment about absolute and relative visual motion into component tasks, it is possible that judgments about motion are made in a holistic fashion. Therefore, at least one dynamic spatial reasoning task should require coordination of time, direction, and motor movement judgments. The act of making a coordinated judgment might itself be a significant source of individual differences and we designed a task that required such judgments.

Path Memory

This task was designed to assess a person's memory for the path of moving objects. On each trial, a small square moved across the computer screen three times. The square followed a parabolic path, starting at the lower left of the computer screen. Either the first and second paths were the same and the third path was different or the second and third paths were the same and the first path was different. The observer indicated which path differed from the second. The computer then reported whether or not the response was correct.

Three parameters determined the paths: the starting height of the parabola, the height of the apex of the parabola, and the horizontal distance from the start of the parabola to the apex. Within each of these dimensions, eight levels of difficulty were established. The easier the level of difficulty, the larger the difference between the unique path and the two identical paths. An adaptive staircasing method was used to measure performance. The level of difficulty was increased when the subject answered two consecutive trials correctly, and decreased when an error was made. A separate staircase was computed for each of the three different methods of changing the parabola. The dependent measure was the average difficulty level of the last two-thirds of the trials in each staircase series.

Extrapolation

This task was intended as a measure of the ability to extrapolate from an observed to an expected path. Three types of curves were presented: a straight line, a sine wave, and a parabola. A portion of the curve was shown on each trial, starting from the left side of the computer screen and ending 41 percent, 52 percent, or 63 percent of the distance across the computer screen. The observer used a joystick to move an arrow up or down along a vertical line on the right side of the computer screen to indicate where the curve would intersect the line. The computer displayed the remaining portion of the curve as soon as the response was made. The dependent measure was the difference between the correct answer and the subject's answer. Separate scores were computed for each of the three types of curves.

Arrival Time for One Object

In this task, the observer had to make an absolute judgment of velocity. On each trial a square moved horizontally from the left side of the computer screen toward a vertical line on the right. One-quarter to one-half of the way across the computer screen, the object disappeared. The observer pressed a key when he or she thought the object would have crossed the line, assuming that it would have continued moving on the same course at the same speed. The dependent measure was the difference in time between the correct answer and the subject's answer. Both absolute differences (accuracy) and signed differences (bias) were computed.

Arrival Time Comparison for Two Objects

This task required judgments of relative speed of motion. On each trial, two different objects were presented, each moving toward its destination at a constant speed. The destinations were displayed as horizontal or vertical lines. A fifth of the way to the destination, the objects disappeared. The subject then reported which object would have arrived at its destination first. The computer informed the subject whether or not the response was correct.

There were five variations of this task. In the first variation, the objects moved in perpendicular paths toward different destinations. In the second variation, the objects moved in perpendicular paths toward the same destination. In the remaining three variations, the objects moved in parallel horizontal paths. In the third variation, the two paths were near each other and the destinations were vertically aligned. In the fourth variation, the two paths were near each other, and the starting locations were vertically aligned. The fifth variation was like the third except that the paths were not near each other; one path was at the top of the computer screen and the other path was at the bottom of the computer screen.

The same staircasing method of measuring performance used in the Path Memory task was used here. Eight levels of difficulty were established by varying the size of the difference in arrival time for the two objects. The staircasing was run separately on the five different variations of the task. The dependent measure was the average level of difficulty for the last two-thirds of the trials.

Arrival Time Comparison for Four Objects

This task measured the ability to deal with relative motion in a somewhat different way. The task was something like guessing the winner of a horse race before the race is completed. The digits 1, 2, 3, and 4 moved horizontally from right to left at individually determined constant speeds. Their destination was a vertical line on the far left of the computer screen. The digits started at varying distances from the line. Half-way to the vertical line, the digits disappeared. If they had continued traveling at the same speed three of them would have arrived at the line at the same time and one would have arrived earlier. The observer indicated which object would have arrrived at the line first.

The same staircasing method was used as in Path Memory and Arrival Time Comparison for Two Objects. There were again eight levels of difficulty, corresponding to the size of the time difference between the arrival time of the first object and the remaining objects. The subject's score was the level of difficulty for the last two-thirds of the trials.

Intercept

This task was designed to measure the ability to combine the extrapolation of both speed and path. The task was like a video game in which a player attempts to "shoot down" a moving object. A small rectangular target moved from left to right at a constant horizontal speed. The target moved along either a horizontal, sine wave, or parabolic trajectory. When the subject pressed a key on the keyboard, a triangular-shaped object (called a "missile") began moving straight upwards at a constant velocity. The subject attempted to time the missile's "launch" so that the missile would collide with the target. The dependent measure was the vertical distance between the missile and the target when the target crossed the missile's path.

GENERAL TESTING PROCEDURES

Each subject performed a total of 19 tests, with the tests distributed over five successive days with two hours of testing per day. The 19 tests represented the 11 computer-administered tasks and 8 paper and pencil tests. For the two-hour testing session on each of the five days of testing the subject received a mixture of computer-administered and paper and pencil tests.

The paper and pencil tests were chosen to represent the major visual-spatial ability factors previously mentioned. There were six tests representing markers for specific visual-spatial abilities. The marker for perceptual speed was the Identical Pictures test from the Educational Testing Service's (ETS) Reference Kit of Cognitive Factors (Ekstrom, French, & Harman, 1979). There were two marker tests for spatial relations ability. One was the two-dimensional spatial relations test from the Primary Mental Abilities Battery (Thurstone, 1965) and the other was a three-dimensional spatial relations test developed by Lansman (1981). The latter was designed as a paper and pencil analog of the mental rotation task developed by Metzler and Shepard (1974). The marker for spatial visualization was the surface devolment space test from the Differential Aptitude Battery (Bennett, Seashore, & Wesman, 1974). The remaining two spatial ability tests in the battery included the spatial orientation test from the Guilford-Zimmerman Aptitude Survey (Guilford & Zimmerman, 1947) and the shape memory test from the ETS Reference Kit. In addition, a vocabulary test from the Nelson-Denny Reading test and a general intelligence test, the Raven's Advanced Progressive Matrixes (Raven, 1962), were included as markers of abilities (gc and gf) outside the spatial domain. For all of the paper and pencil tests, the dependent measure was the number correct with a correction for guessing where appropiate.

OVERVIEW OF RESULTS

The 11 computer-administered tasks produced a wide variety and range of individual performance scores. In the present context it is not possible to provide a complete discussion of the descriptive data or the analytic details with regard to all the various performance measures. For those interested in such information, the report by Hunt et al. (1986) should be consulted. What follows is a summary of the most important results as they relate to two key issues. The first major issue is whether the data support the conclusion that computer-administered, "theory-based" tasks of static spatial reasoning can be used to replace and augment current forms of spatial ability assessment. The second major issue is whether the data support the conclusion that computer-administered tests of dynamic spatial reasoning define an ability separate from that assessed by computer-based or conventional tests of static spatial reasoning. In the next two sections we consider each of these issues in turn as well as subsidiary issues embedded within each larger issue.

ISSUE 1—THE ASSESSMENT OF STATIC SPATIAL REASONING

To address the issue of whether computer-based tasks of static spatial reasoning can replace and augment current paper and pencil tests of static spatial reasoning, a series of subsidiary issues must be addressed. The first such issue is whether the computer-based tasks and measures behave in a principled manner and have substantial reliability. Each of our five computer-based tasks of static spatial reasoning had a systematically designed problem set permitting internal validation of a task theory of processing with simultaneous derivation of overall latency and accuracy measures as well as more refined component process measures. The group and individual subject data revealed that performance in all five tasks was consistent with general expectations about within-task processing.

Table 8.1 is a summary of overall latency and accuracy scores obtained within each task and the reliability of each measure. In the perceptual comparison task, accuracy was generally high and response latency was a systematic function of stimulus complexity and degree of similarity between stimuli. In the mental rotation task, accuracy was also generally high and response latency was a systematic function of angular disparity between stimuli. In the surface development task, latency and accuracy were systematic functions of the number of mental folding operations. In the integration of detail task, integration latency was a systematic function of the number of elements to be integrated to form a composite image. In the adding detail task, decision latency and accuracy were a systematic function of the number of details in the final image. As can be seen in Table 8.1, the overall latency and accuracy measures had substantial reliabilities, with the majority of reliability coefficients above .90. Within each task we also derived various component process latency measures. Examples include the slope and intercept of the mental rotation function and the slope of the function relating integration latency to the number of elements to be integrated. These component process latency measures also had substantial reliability and the average reliabilities of such within-task measures ranged from .67 to 91.

Given that the various overall performance and component measures have substantial reliability, then the second issue is whether these measures encompass the variance produced by performance measures obtained from

Table 8.1
Reliabilities of Computer-Controlled Static Task Performance Measures

	Measure	
Task	Mean Accuracy	Mean Latency
Perceptual Comparison	.93	.98
Mental Rotation	.97	.99
Surface Development	.95	.95
Integrating Detail	.75	.93
Adding Detail	.60	.95

the paper and pencil tests. The analytic strategy for addressing this issue in-volved several stages. The first stage was to reduce the variables to be analyzed to a manageable number. To do so, separate factor analyses were conducted for all variables within each computer-administered task, using an orthogonal factor analysis followed by varimax rotation (Mulaik, 1972). At most three factors were extracted from each set of within-task measures and these typically represented separate latency and accuracy factors. Only the best marker for each of these factors was retained for further analyses and these measured are shown in Table 8.2. All scores were scaled so that high scores reflect better performance.

The second stage of analysis was to conduct separate factor analyses of the paper and pencil task measures and the computer-administered task measures to determine the dimensionality within each domain of perfor-mance. Recall, from the introductory discussion, that the paper and pencil tests would be expected to show either a two- or a three-dimensional factor pattern. Table 8.3 presents the results of an orthogonal factor analysis of the paper and pencil tests followed by a varimax rotation. As expected, the analysis identified two factors with eigenvalues greater than 1. The two-dimensional space accounted for 54 percent of the variance between measures. The first factor is closely related to the two rotation tests, but is also associated with all the other primary measures of spatial ability. The Raven Matrix test has only a small loading on this factor, and the vocabulary test is virtually orthogonal to this factor. The second factor is identified by relatively high loadings on the more complex, power-oriented spatial visualization tests, and by a very high loading on the Raven Matrix test. This test depends both on spatial and abstract reasoning (Hunt, 1974). The factor analysis results indicate that the paper and pencil test scores

Table 8.2
Computer-Controlled Static Task Performance Measures

Task	Measure	Abbreviation
Perceptual Comparison	Mean Response Latency	PC-LAT
	Mean Accuracy	PC-ACC
Mental Rotation	Mean Response Latency	ROT-LAT
	Mean Accuracy	ROT-ACC
Surface Development	Mean Response Latency	SD-LAT
	Mean Accuracy	SD-ACC
Integrating Detail	Mean Integration Latency	ID-ILAT
	Mean Response Latency	ID-RLAT
	Mean Accuracy	ID-ACC
Adding Detail	Mean Add Latency	AD-ALAT
	Mean Response Latency	AD-RLAT
	Mean Accuracy	AD-ACC

obtained in the current study are distributed much as one would expect them to be given previous factor analytic research.

A similar factor analysis was conducted for the selected measures from the computer-administered static tasks. The factor matrix is shown in Table 8.4. The first factor shows the highest loadings for measures of processing accuracy. The second factor generally has positive loadings for the latency measures. The third factor is associated with the latency measures from the adding detail task. The fourth factor appears to be primarily associated with the measures from the perceptual comparison task and reflects a speed accuracy trade-off in performance within this task.

To examine these findings further, an oblique (oblimin) factor analysis was conducted. This method was chosen because the psychological processes just described would predict nonindependence between measures of speed and accuracy, both within and across tests. Table 8.5 presents the results of the oblique factor analysis, and Table 8.6 shows the correlations between the factors. The pattern in Table 8.5 is similar to that in Table 8.4, except that Factors I and II are reversed, as are Factors III and IV. Factor I

Table 8.3
Factor Matrix for Paper and Pencil Tests

Test Score	Factor I	Factor II
Identical Pictures	.59	.06
PMA Rotation	.70	.18
3-D Rotation	.62	.25
DAT Surface Development	.53	.53
GZ Orientation	.42	.58
Spatial Memory	.16	.34
Raven Matrix	.28	.73
Vocabulary	-.01	.25

Table 8.4
Factor Matrix for Computer-Controlled Static Tasks

Measure	Factor I	Factor II	Factor III	Factor IV
PC-LAT*				.80
ROT-LAT		.38		.41
SD-LAT		.60		
ID-ILAT		.73		
ID-RLAT		.66		
AD-ALAT			.84	
AD-RLAT		.30	.67	
PC-ACC	.30			-.62
ROT-ACC	.42			
SD-ACC	.75			
ID-ACC	.69			
AD-ACC	.46			

* Cells with values below .30 have been left empty for ease of viewing.

Table 8.5
Oblique Solution Factor Matrix for Computer-Controlled Static Tasks

Measure	Factor I	Factor II	Factor III	Factor IV
PC-LAT*	.33		-.82	
ROT-LAT	.44		-.48	
SD-LAT	.62			
ID-ILAT	.75	-.31		.35
ID-RLAT	.69			.31
AD-ALAT				.86
AD-RLAT	.42		-.35	.74
PC-ACC		.31	.65	-.36
ROT-ACC		.41		
SD-ACC		.76		
ID-ACC		.70		
AD-ACC		.49		-.35

* Cells with values below .30 have been left empty for ease of viewing.

Table 8.6
Correlations between Oblique Factors

	Factor II	Factor III	Factor IV
Factor I	-.07	-.37	.33
Factor II		.01	-.23
Factor III			-.29

has high loadings for latency measures, and only small loadings for accuracy measures. Factor II, conversely, is characterized by high loadings for accuracy measures. The two factors are essentially uncorrelated. Superimposed on this pattern, however, are factors III and IV. Factor II is a bipolar factor for the perceptual comparison task, again suggesting a strong speed-accuracy trade-off across individuals for this task. (This is consistent with experimental results examining the task in detail and the within-task factor analysis of performance measures.) Factor IV is a similar, somewhat less strongly defined speed-accuracy trade-off for the adding detail and integrating details task.

The conclusion that can be drawn from these separate factor analyses within the domain of paper and pencil tests and the domain of computer-based static display tasks is that the computer-controlled static spatial reasoning tasks appear to encompass the same spatial relations and visualization factors as the paper and pencil tests while at the same time offering the potential for distinguishing between speed and accuracy of processing as well as speed-accuracy trade-offs within tasks.

The third stage of analysis was to determine more precisely whether or not the two domains of tests tap the same psychological abilities. To determine whether the individual variation captured by the paper and pencil tests is embedded in the static task measures, a canonical correlation analysis was conducted. A canonical correlation locates spaces of common variance embedded within each of the two domains (paper and pencil and static tasks), and then computes the (maximized) correlations between the dimensions of each of the spaces (Cohen & Cohen, 1975). At most two canonical correlates should be constructed, because the common variance in the lower order space (the paper and pencil tests) is apparently two-dimensional (see the above discussion of the factor analysis within the domain of paper and pencil tests). As expected, two canonical correlates were extracted and they accounted for 90 percent of the variance. The first canonical correlation was .78 and the second was .54, both statistically reliable at p < .01. Therefore, it can be concluded that a substantial portion of the common variance in the paper and pencil domain is embedded within the (three- or four-dimensional) common variance of the computer-controlled static tasks.

The results support the conclusion that our theory-based, computer-administered static spatial reasoning tasks can be used to replace and augment current paper and pencil procedures for the assessment of spatial ability. The computer-based tasks and measures (1) behave in predictable ways, (2) have high reliability, (3) encompass the variance contained in traditional paper and pencil tests, and (4) seem to provide measures of unique variance associated with speed versus accuracy of processing.

ISSUE 2—STATIC VERSUS DYNAMIC SPATIAL ABILITIES

The issue of whether the computer-administered tasks of dynamic spatial reasoning assess an ability separate from the abilities assesed by static spatial tasks also required that a series of subsidiary issues be addressed. The first such issue was whether performance in the computer-administered dynamic tasks was reliable. Table 8.7 shows the reliabilities of the different performance measures derived in each of the six dynamic tasks. In three of the tasks, path memory, arrival time comparison for two objects, and arrival time comparison for four objects, a staircasing method was used to adjust problem difficulty and the measure of performance was the average difficulty level for the last two-thirds of the problems presented. As shown in Table 8.7, the reliability of the average difficulty level was moderate for two of these tasks and low for the third task. We believe that these somewhat disappointing reliabilities reflect an insufficient number of trials for adequate assessment of difficulty level in these comparative judgment tasks. The measures derived for the other three dynamic tasks, extrapolation, intercept, and arrival time for one object, had adequate to very high reliability levels.

The next issue to be considered was the dimensionality of the space defined by the measures from the dynamic spatial processing tasks. An orthogonal factor analysis was computed using the measures shown in Table 8.7. The factor analysis before rotation indicated that a two-factor solution was required, with 44 percent of the common variance between tests located in a two-dimensional space. This solution was made simpler by a varimax rotation, which identified two factors. These are shown in Table 8.8. The first factor, which accounted for 71 percent of the common (two-space)

Table 8.7
Reliability of Dynamic Task Performance Measures

Task	Measure	Reliability
Path Memory	Difficulty Level	.50
Extrapolation	Weighted Average Error	.81
Arrival Time (1 Object)	Absolute Error	.99
	Bias	.99
Arrival Time (2 Objects)	Difficulty Level	.63
Arrival Time (4 Objects)	Difficulty Level	.30
Intercept	Weighted Average Error	.79

variance, was marked by high loadings on the arrival time comparison for two- and four-object tasks and the intercept task. The second factor was associated primarily with the extrapolation task. Note that the extrapolation task can be solved without considering its dynamic aspects, since at the time the examinee must respond, a static picture is present and the information in the picture is sufficient to define the correct answer. The commonalities of the memory for path task and the measure in the time of arrival-one object task were very low (less than .1 in all cases), suggesting that these measures do not tap processes associated with the other tasks.

Given the preceding results, canonical correlation analysis was again used to explore the connection between the dynamic tasks and the other two task domains. In each case we would expect at most two canonical correlates, because of the low dimensionality of the dynamic tasks. There were two canonical correlates between the static and dynamic task performance measures, with correlation values of .59 and .46, < .01 and .05, respectively. These accounted for 56 percent of the variance. There was only a single canonical correlate connecting the paper and pencil and dynamic task measures, with a canonical correlation of .60, p < .001, which accounted for only 36 percent of the variance. Although these two sets of correlations involving the dynamic task measures are statistically reliable, they are substantially below the correlations that were found between the paper and

Table 8.8
Factor Matrix for Computer-Controlled Dynamic Tasks

Task/Measure	Factor I	Factor II
Path Memory	.26	
Extrapolation		.74
Arrival Time		
Error		
Bias		
Arrival Time (2 Objects)	.59	
Arrival Time (4 Objects)	.46	
Intercept	.55	

* Cells with values below .25 have been left empty.

pencil and static task performance measures. This indicates that the dynamic tasks tap processes that are correlated with, but not identical to, the processes required to execute the various static tasks.

As a further test of the hypothesis that the dynamic tasks define a separate ability, a confirmatory factor analysis (Joreskog & Sorbom, 1979) was conducted, analyzing the common covariance of the static and the dynamic computer-controlled tasks. In this analysis, attention was restricted to measures from the intercept and the arrival time comparison tasks, as they gave the clearest indication of being good measures of a dynamic spatial factor. Three principles were used to construct the hypothesized factor structure. They were: (1) Three factors were assumed, a latency factor, an accuracy factor, and an "ability to deal with dynamic motion" factor. The first factor was defined by latency measures from the static tasks, the second by accuracy measures from the static tasks, and the third by the dynamic task measures. (2) Correlations between the three factors were permitted. (3) The "task-specific" (residual) components of each measure taken from the same task were assumed to correlate. The rationale behind this is that these measures are based on different analyses of the same physical response. Any event in time that affects a response, but that is logically irrelevant to the test situation (for example, the examinee's attention being momentarily distracted) should affect all measures based on the same response.

The path diagram and parameter estimates for this model are shown in Figure 8.1 and the chi square and other degree of fit statistics are shown in Table 8.9. The chi square values and the chi square divided by degree of freedom ratio are sufficiently high to be of concern, but the goodness of fit indexes and the root mean square values are comparable to those obtained for the within-domain models. Most importantly, examination of the details of the deviations from a perfect fit indicated that the problems were in the relations the various static test measures, and not in the relation between the static and dynamic tests.

Further belief in the need for a separate dynamic movement processing factor was obtained by conducting a similar confirmatory factor analysis in which the dynamic process factor was eliminated, and the dynamic tasks were related to the latency and accuracy factors. Although this model has fewer degrees of freedom than does the model of Figure 8.1, the chi square value was higher. While a direct comparison of the models is not possible, because one does not fully contain the parameters of the other, a model with more parameters and worse fit is hardly preferable to one with fewer parameters and better fit.

We conclude that there is strong evidence for a dynamic spatial reasoning processing factor that is separate from the static spatial reasoning factors previously identified. A further examination of Figure 8.1 shows the nature of the relation between the static and dynamic spatial processing traits. Because the model is based upon an analysis of a correlation matrix, the

Figure 8.1
Path Diagram and Parameter Estimates for the Model with Two Static and One Dynamic Spatial
Processing Factors

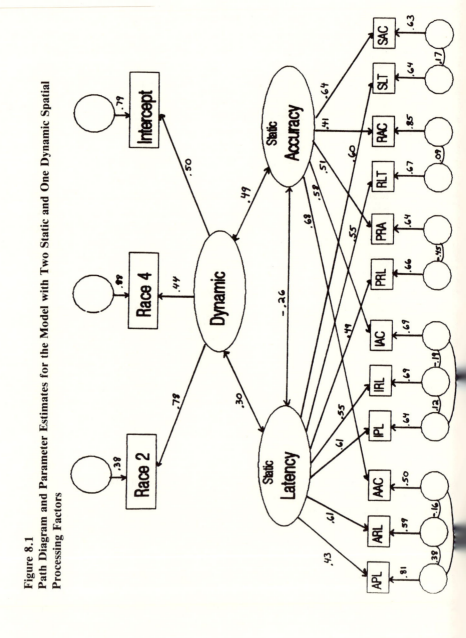

Table 8.9
Indexes of Fit for Confirmatory Factor Analysis

Chi Square	163.07
Degrees of Freedom	78
Probability Level	<.001
Chi Square/D of F	2.09
Joreskog Index	.92
RMSD	.10

variance of all latent traits was set to one. Thus, the path coefficients can be interpreted as coefficients between standardized variables. There is a small, negative relationship between latency and accuracy on the static tasks which was not statistically reliable. The latent trait underlying dynamic spatial ability is reliably related to both the latency trait and the accuraccy trait. Both of these coefficients were reliably greater than zero.

SUMMARY AND CONCLUSIONS

In this chapter we have reviewed some of the results of a project focusing on advancements in testing as applied to the domain of spatial abilities. The research is a first attempt to evaluate the utility of combining information processing theories and componential analyses of human abilities with computer technology. At the same time, it attempts to go beyond current paper and pencil and componential assessment of reasoning about static spatial information by exploring the possibility of a separate ability for reasoning about dynamic spatial relations. The latter is only feasible in the context of computerized testing.

The general conclusions from our research project can be stated quite simply. The results of various analyses indicated that the individual variation in static spatial reasoning captured by conventional paper and pencil tests can also be captured by computer-controlled analogs of these tests. The computer-based tasks are preferable to the paper and pencil tests for several reasons. First, they provide highly reliable general and specific measures of the speed and accuracy of spatial reasoning within individual tasks. The specific componential measures are related to theories of spatial information processing and thus can be understood within such a theoretical context which is separate from the multivariate context of psychometric theories of spatial ability. In addition, the computer-based static spatial reasoning tasks provide ways of allowing for and measuring individual differences in speed-accuracy trade-offs. Very little research has been done to explore this issue, although speed-accuracy trade-offs have been shown to be related to age and personality factors.

Our results also indicate quite strongly that the ability to deal with moving elements and dynamic spatial relations is separate from the abilities associated with reasoning about static spatial information. The dynamic spatial ability factor we have identified is largely related to dealing with relative visual motion. This is the type of judgment required by the two arrival time comparison tasks and the intercept task. Our test battery also included tests of the ability to make absolute judgments about motion, path extrapolation, and path memory. Although these tests were generally reliable they do not appear to play a major role in the definition of the latent trait underlying performance in the dynamic spatial reasoning tasks. Perhaps this reflects nothing more than the need for additional tests in a larger battery to identify other dynamic spatial processing factors. Further research will be required to examine this issue.

The present results regarding existence of a dynamic spatial reasoning factor hold both theoretical and practical promise. From a theoretical perspective, there are numerous issues to be addressed about the nature of representation and processing of dynamically changing spatial information. This includes how such representations and processes relate to extant theories of imagery and static spatial processing. We have only begun to scratch the surface with respect to these issues.

From a practical perspective, there is the distinct possibility that the ability to represent and reason about dynamically changing spatial relations may better predict real-world performances that intuitively seem to require these abilities. Examples include machinery operation and piloting. Of further significance is the fact that this ability seems separate from those assessed by conventional spatial ability tests. Our results do not address the question of the utility of the computer-controlled tests of either static or dynamic spatial reasoning as predictors of performance in situations outside the laboratory. As noted at the beginning of this chapter, this is an area of considerable interest and will have to be the topic of new research. The work we have completed serves to demonstrate the practicality of componential, computer-based approaches to spatial ability testing and is offered as a necessary precursor to such validation research.

REFERENCES

Alderton, D., and Pellegrino, J. W. (1984). *Analysis of mental paper folding.* Unpublished manuscript, University of California, Santa Barbara, CA.

Bennett, G. K., Seashore, H. G., and Wesman, A. G. (1974). *Differential Aptitude Test.* New York: The Psychological Corporation.

Carroll, J. B. (1982). The measurement of intelligence. In R. J. Sternberg (Ed.), *Handbook of human intelligence* (pp. 29–120). Cambridge: Cambridge University Press.

Cohen, J., and Cohen, P. (1975). *Applied multiple regression and correlational analysis for the behavioral sciences.* Hillsdale, NJ: Erlbaum.

Cooper, L. A. (1976). Individual differences in visual comparison processes. *Perception & Psychophysics, 19,* 433–444.

Ekstrom, R. B., French, J. W., and Harman, H. H. (1979). Cognitive factors: Their identification and replication. *Multivariate Behavioral Research Monographs* (no. 97-2), 1–84.

Guilford, J. P., and Zimmerman, W. S. (1947). *Guilford-Zimmerman aptitude survey.* Orange, CA: Sheridan Psychological Service.

Hunt, E. B. (1974). Quote the Raven? Nevermore! In L. Gregg (Ed.), *Knowledge-cognition.* Hillsdale, NJ: Erlbaum.

Hunt, E., and Pellegrino, J. (1985). Using interactive computing to expand intelligence testing. *Intelligence, 9,* 207–236.

Hunt, E., Pellegrino, J. W., Abate, R., Alderton, D., Farr, S., Frick, R., and McDonald, T. (1986). *Computer controlled testing of spatial-visual ability,* (NPRDC Technical Report). San Diego, CA.

Joreskog, K., and Sorbom, D. (1979). *Advances in factor analysis and structural equation models.* Cambridge, MA: Abt Books.

Kosslyn, S. M. (1980). *Image and Mind.* Cambridge, MA: Harvard University Press.

Lansman, M. (1981). Ability factors and the speed of information processing. In M. P. Friedman, J. P. Das, and N. O'Connor (Eds.), *Intelligence and learning* (pp. 441–457). New York: Plenum Press.

Likert, R., and Quasha, W. H. (1971). *Revised Minnesota paper foram board test (Series AA).* New York: The Psychological Corporation.

Lohman, D. F. (1979). *Spatial ability: A review and reanalysis of the correlational literature* (Report No. 8, Aptitude Research Project). Palo Alto, CA: Stanford University.

Lohman, D., Pellegrino, J. W., Alderton, D., and Regian, J. W. (1986). Dimensions and components of individual differences in spatial abilities. In S. Irvine and S. Newstead (Eds.), *Intelligence and cognition: Contemporary frames of reference* (pp. 253–312). Dordrecht, Netherlands: Martinus Nijhoff.

McGee, M. G. (1979). Human spatial abilities: Psychometric studies and environmental, genetic, hormonal, and neurological influences. *Psychological Bulletin, 86,* 889–918.

Metzler, J., and Shepard, R. N. (1974). Transformational studies of the internal representations of three-dimensional objects. In R. Solso (Ed.), *Theories in cognitive psychology: The Loyola Symposium* (pp. 147–201). Hillsdale, NJ: Erlbaum.

Mulaik, S. A. (1972). *The foundations of factor analysis.* San Francisco: McGraw-Hill.

Mumaw, R. J., and Pellegrino, J. W. (1984). Individual differences in complex spatial processing. *Journal of Educational Psychology, 76,* 920–939.

Mumaw, R. J., Pellegrino, J. W., Kail, R., and Carter, P. (1984). Different slopes for different folks: Process analyses of spatial aptitude. *Memory & Cognition, 12,* 515–521.

Pachella, R. G. (1974). The interpretation of reaction time in information-processing research. In B. H. Kantowitz (Ed.), *Human information processing: Tutorials in performance and cognition* (pp. 41–82). Hillsdale, NJ: Erlbaum.

Pellegrino, J. W. (1984). *Information processing and intellectual ability.* Paper presented at the AERA annual meeting, New Orleans, LA.

Pellegrino, J. W., Alderton, D., and Shute, V. J. (1984). Understanding spatial ability. *Educational Psychologist, 19,* 239–253.

Pellegrino, J. W., and Kail, R. (1982). Process analyses of spatial aptitude. In R. Sternberg (Ed.), *Advances in the psychology of human intelligence* (Vol. 1, pp. 311–366). Hillsdale, NJ: Erlbaum.

Pellegrino, J. W., Mumaw, R. J., and Shute, V. J. (1985). Analyses of spatial aptitude and expertise. In S. E. Embretson (Ed.), *Test design: Developments in psychology and psychometrics* (pp. 45–76). New York: Academic Press.

Poltrock, S. E., and Brown, P. (1984). Individual differences in visual imagery and spatial ability. *Intelligence, 8,* 93–138.

Raven, J. C. (1962). *Advanced progressive matrices set II.* London: H. K. Lewis.

Shepard, R. N., and Cooper, L. A. (1982). *Mental images and their transformations.* Cambridge, MA: MIT Press.

Shepard, R. N., and Feng, C. (1972). A chronometric study of mental paper folding. *Cognitive Psychology, 3,* 228–243.

Sternberg, R. J. (Ed.). (1985a). *Human abilities: An information processing approach.* New York: Freeman.

———. (1985b). Cognitive approaches to intelligence. In B. B. Wolman (Ed.), *Handbook of intelligence: Theories measurements and applications* (pp. 59–118). New York: John Wiley.

Thurstone, L. L., and Thurstone, T. G. (1949). *SRA primary abilities.* Chicago: Science Research Associates.

Thurstone, T. G. (1965). *Primary mental abilities.* Chicago: SRA.

9   Situationist Charges versus Personologist Defenses and the Issue of Skills

Luc Bovens
Arnold Böhrer

In this chapter we intend to discuss a few theoretical issues in personality theory and in the methodology of personality measurement. On the background of this discussion we will then present some of the personality tests that were developed (at least partly) within the Section for Psychological Research in the Center for Recruitment and Selection of the Belgian Army.[1]

SITUATIONISM VERSUS PERSONALITY THEORY

Situationist Charges and Personologist Defenses

The first theoretical issue that we are planning to consider is whether the very concept of personality (1) can play a useful role in contemporary psychology or (2) is merely an aggravating legacy from *folkpsychology* (that

We are grateful to the numerous people who have in some way contributed to the research on the free-format self-description method (VZM). We would especially like to mention P. De Boeck, R. Biesmans, S. Van Den Broucke, and the Laboratory for Personality Research in the Dept. of Psychology in the K. U. Leuven. For research on the ROMAT and the RS1C we respectively thank Ph. Schotte and P. Vermeulen.

is, the common art of explaining or predicting actions, emotions, etc., which people generally practice in living their lives).[2] Here is how the debate runs. Let's start off with a caricature of the latter position, which we promise to label "extreme situationism." The extreme situationist claims that all psychological explanation that makes reference to persistent personality traits:

(b_1) is merely ad hoc and

(b_2) cannot be but unsuccessful, since there simply is no constancy in a person's behavior when one is confronted with differing environments.

Let us at first consider objection (b_1). Molière ridiculed the science of his age in a play which stages a medical student proclaiming solemnly that opium puts a person to sleep due to its *virtus dormitiva* or somniferous power (Molière, 1947, p. 196). Now the extreme situationist claims that this is precisely the kind of situation that the personality theorist is in, say, in explaining hysteric outbursts in reference to a neurotic character and that all personologist theorizing actually has the same ad hoc character to it.

Consider now objection (b_2). In trying to explain or predict some action, folkpsychology does pay due attention to person-variables ("What sort of person are we dealing with?"). Now extreme situationists claim that this folkpsychological mode of explanation or prediction should not be adopted in science, since all action is, as a matter of fact, determined only by situation-variables. Across sufficiently differing environments there simply is not constancy in a person's behavior, and consequently, person-variables lack all explanatory or predictive power. Thus, if the situationist objections hold, the personality theorist would even be worse off than Molière's scientist. Reference to a somniferous power is not very informative, though it is reasonable to assume that opium will have the same power tomorrow as today. The personality theorist, on the other hand, cannot even assume that the ad hoc features which he calls in have any persistence over time.

Now, what kind of response does the personality theorist offer to these extreme situationist charges in defense of his or her project? The personality theorist believes that personality traits are a necessary component in psychological theorizing. The personality theorist, in granting that the situationist objections may more or less hold against folkpsychological traits, believes that (a_1) situation-variables, taken by themselves, are insufficient to explain or predict actions and (a_2) that the situationist objections can be overcome if we are sufficiently careful in selecting the set of personality traits that are to play a role in psychological theory.

What sort of traits then are to play a role in proper psychology? Within personality theory itself, there are widely divergent opinions concerning what should count as an adequate personality feature in psychological theory. Consider the set of categories in our common practice of personality trait ascription, that is, the set containing descriptive terms like "honest," "friendly," "outgoing," etc. Now, this set of folkpsychological terms is apparently inadequate for doing proper psychology. But how should we adjust this set?

There are two such psychological projects which we would like to discuss within this chapter, namely, the construction, on the one hand, of a set of situation-specific traits and, on the other hand, of a set of five personality factors. The first project specifically aims at insuring the *constancy* of personality traits, though we believe the price for this is that a psychological explanation which makes reference to such situation-specific traits even has more of an ad hoc character than folkpsychological explanations. Conversely, the second project reduces the ad hoc character of psychological explanations, though the price for this may well be that such factor has even less constancy than folkpsychological features.

Let us now take a close look at both projects. The first project is connected with the name of Lawrence A. Pervin. Pervin (1976, 1977) asks his subjects to:

1. list some significant situation in their current lives;
2. describe these situations;
3. describe how they feel about each situation;
4. describe how they behaved in each situation.

He then isolates four sets of factors within the set of situations for each particular individual. In the first set similar descriptions load on the respective factors, in the second set similar feelings, in the third set similar behaviors, and in the fourth set similar combinations of descriptions, feelings, and behavior. Now this analysis yields four lists of situation-specific trait decriptions for each person. These respective lists contain the following types of statements.

List 1: Person X is the kind of person who typically *describes* y-related situations as z_1.

List 2: Person X is the kind of person who typically *feels* z_2 about y-related situations.

List 3: Person X is the kind of person who typically *reacts* z_3 to y-related situations.

List 4: Person X is the kind of person who typically, for y-related situations, describes them as z_1, feels z_2 about them, *and* reacts in a z_3 way to them.

The set of factors y includes things like school, home, friends, etc.; z_1s are adjectives descriptive of situations, such as comfortable, frustrating, demanding, etc.; z_2s are adjectives denoting emotions, like happy, nervous, scared, etc.; z_3s are adjectives denoting types of behaviors, like responsible, friendly, irascible, etc.

Now this set of lists will yield a very detailed and structured description of what kind of person X is. We can also generalize concerning the rate of occurrence of certain descriptions, feelings, and behaviors, or concerning particular interrelations between descriptions, feelings, and behaviors. These empirical data show that for a particular person some traits are highly general, that is, hold across a broad set of situation-bound, that is, hold only for a small set of situations. Such a situation-specific trait description, that is, a trait description that includes reference to the range of situations to which the trait applies, is Pervin's alternative to folkpsychological trait description. And this mode of trait description, Pervin argues, can (at least partly) avoid the situationist objection that traits have no constancy. Now this indeed seems to be a reasonable claim. Most people certainly do adjust their styles to different types of situations. Consequently, Pervin's situation-specific trait descriptions will do better on the issue of constancy than our much less situation-specific folkpsychological trait descriptions.

On the other hand, we do have serious doubts about the explanatory value of such traits. It seems to us that in explaining an action, the more situation-specific the explanatory trait becomes, the more we are dissatisfied with its ad hoc character. Here is an example: We are interested to know why Mary blew up at her coach during soccer practice when the coach told Mary she expected more commitment from her. Now following Pervin's model we may say that Mary is the kind of person who reacts irritably (z_3) in sports (y) when she feels personally attacked (z_2) in a threatening situation (z_1). But is this a genuinely satisfactory explanation? It seems to us that what we would really want to know is *why* Mary is precisely such a kind of person. Now answering this question forces us to use more general, and thus less situation-specific trait descriptions. We may want to say that Mary is a very quarrelsome personality, who would construe any situation as being threatening just in order to pick a fight. Or maybe Mary is just in general a very sensitive character, who is very much affected by remarks concerning her performance. Or maybe Mary is very competitive, such that her whole life turns around sports: consequently she takes such comments by her coach very personally. Now we think it is these sorts of explanations which would genuinely satisfy us. But then of course, with more general personality-traits, the problem of constancy again becomes more acute.

Let us now turn to the 5-personality-factor theory. Tupes and Christal (1961) presented their subjects with a checklist with 35 pairs of bipolar trait-names (silent vs. talkative, depressed vs. cheerful etc.) This set of pairs they considered to be representative of the full set of trait-names in the English language. The subjects in their study were asked to check the adjectives in each pair which they considered to be most descriptive of their personality. A factor analysis yielded five *factors* or sets of intercorrelated pairs of adjectives, which were labeled "surgency," "agreeableness," "independability," "emotional stability," and "culture." Now much research has been done on fine-tuning and redefining these categories, as well as on identifying the same set of factors in data from other measurement-techniques (questionnaires, peer-ratings) (for example, McCrae et al., 1985a, 1985b, 1986, 1987). McCrae and Costa claim that these five factors (or actually, a close variant) provide for the basic structure of personality and that any model of personality *must* include at least these five factors. Whether it should include more depends on the scope one wants to assign to the concept of personality, for example, the five-factor model does not include cognitive abilities or elements of self-concept like body image or social identity (McCrae & Costa, 1986, p. 22). Now this project reduces the set of folkpsychological concepts to an extremely small set. What are the consequences if we decide to do proper psychology with this particular set? On the negative side, it seems to me that this project is more vulnerable to situationist charges of lack of constancy. Consider a person who scores on average on agreeableness. The adjective pairs "suspicious vs. trustful" and "cool, aloof vs. attentative to people" load on the factor of agreeableness. But now imagine this person is very agreeable when it comes to being sensitive to other persons' needs, while not at all agreeable when it comes to trusting other people if it is sensible to do so. In this particular case, the ascription of the feature "being agreeable" in five-personality-factor theory would display less constancy than the folkpsychological ascriptions "being suspicious" or "being attentive to people." On the positive side, it seems that if it were the case that we could construct models in which the dependent variable is some type of action and the independent variables includes a combination of scores on the five-factor scale (plus a set of situation-variables), this would provide for an explanation which supersedes the charges of the ad hoc character of explanations involving personality-features. Situationist charges of ad hoc-ness would become obsolete vis-à-vis explanations invoking general statements of the form "If some person scoring such and such on surgency, agreeableness . . . is placed in ____ circumstances, she is likely to do ____."

Now, what can be concluded from this discussion? We have argued that extending the set of folkpsychological trait names by making them situation-specific may alleviate the problem of constancy while intensifying the problem of the ad hoc character of the explanations thus obtained. On the other hand, reducing the set of folkpsychological trait names to five factors may alleviate the problem of the ad hoc character of the explanations thus obtained, while intensifying the problem of constancy. Now it may thus well be the case that our common folkpsychological trait names provide for an optimal balance between both alternatives. The wisest route to take for proper psychology may thus be to work with our common folkpsychological trait names or close variants in constructing explanatory models. And this, I believe, is what most often happens in actual psychological theorizing anyway.

An Application in Personality Testing: The VZM

The concrete problem we were facing was the development of a personality test for the purpose of selection within a military setting. Now what can we learn from our theoretical reflections as to what features are desirable for a personality test with this particular purpose? In the first place we are interested in a few traits that our folkpsychological intuitions tell us might be good predictors for a successful military career, namely, *leadership* and *creativity*. Furthermore, we are also interested in sketching an overall picture of the test taker's personality. For this purpose we decided it would be desirable to collect scores on a more recent variant of Tupes and Christal's five-personality factors, involving *extroversion, agreeableness, conscientiousness, neuroticism,* and *general culture* (McCrae & Costa 1985a, 1985b; McCrae et al., 1986, 1987). But if we decide to adopt this five-factor model, we must find some way to (more or less) bypass the situationist charges of no constancy. Pervin's research provides for empirical evidence that, while some traits are highly situation-bound, other traits are quite persistent across situations for particular persons. Now it is precisely this last set of traits that we are interested in, in sketching an overall picture of our test takers. But how can we get at precisely these traits? We assumed that, if a person would have a chance to describe herself by means of a list of adjectives, she would be most struck precisely by these character features that affect her life across differing environments. Now, if this assumption holds, free self-description would thus be an appropiate technique to get at more or less constant personality traits.[3] A particular problem of a self-descriptive method in a context of selection is that the test takers will attempt to present an overly positive picture of themselves. In order to check for this tendency we have added one control-dimension, *social desirability*. In order to make our results interpersonally comparable it would be desirable to assign standardized scores on all these dimensions. But now how can we translate such widely diverging lists of self-descriptions into scores on the particular dimensions which we are interested in, namely, *leadership, creativity,* the five-personality factors, and *social desirability*? In order to solve this problem we have worked out a test procedure, the VZM or *V*rije *Z*elfbeschrijvings*m*ethode (the Free-Format Self-Description Method) which closely resembles Potkay and Allen's Adjective Generation Test (AGT) (1973).

The subjects in our test are asked to describe their personality by means of ten adjectives. These adjectives are then looked up in a scoring list. The scoring list contains a large set of "expressions," that is, nouns, adjectives, and descriptive phrases, which might possibly describe a personality feature, with matching scores assigned by groups of judges on each of the eight dimensions. The subject's overall raw score per dimension is then calculated by adding the respective scores for each adjective. These raw scores are then converted and standardized into an 11-point scale with fixed mean and fixed standard deviation. In a last step, these standardized scores are plotted on a graph, which provides for a personality profile of the person in question.

Let us now consider some details. The expressions in this scoring list were compiled from a set of responses from some pilot studies involving more than 3,000 subjects. The scores were obtained by having groups of ten judges rate some hypothetical person who would adopt the expression in question to describe himself, on *social desirability*, on the five factors, on *leadership*, and on *creativity*. If the judge thought the hypothetical person would (not) have the feature in question to a large degree (at all), a score of +2 (−2) was assigned to the particular expression. Intermediate scores were available of +1 and −1, and a score of zero was assigned if the expression did not allow for a judgment on the feature in question. Scores were added over the ten judges and an overall score (between +20 and −20) was assigned for each expression per dimension. Here is an example:

	Ambitious	Persistent
Social Desirability	7	7
Extroversion	5	3
Agreeableness	−2	−1
Conscientiousness	3	9
Neuroticism	−2	−5
General Culture	7	3
Leadership	13	10
Creativity	1	2

To improve the uniformity of scoring in the judging task we attached some descriptive phrases to each dimension (apart from *social desirability*). For example, for *creativity* we added the following specifications: "inventive, imaginative, breaks through old patterns of thought, approaches issues from different angles and comes up with new ideas."

The subjects are encouraged to write down ten adjectives. Some subjects, however, choose expressions which are not adjectives and some hand in lists of less than ten responses. In the scoring list we have included expressions other than adjectives to allow for the former possibility. Furthermore, if a response does not occur in our list, the test examiner may replace this response with a synonym which does occur in the list. (We know from current research that only 10 percent of the responses was not included in the Dutch list.) If no synonym can be found, the response may simply be dropped from the list. (Research has shown that dealing with the problem of responses which do not occur in the list of expressions by either dropping all such responses, searching for *all* such responses, or setting a group of ten judges to work at rating the particular responses did not yield significantly different results. We have thus opted for a method which combines the ease of the first method and the frugality of the second, while avoiding the complexity of the last method.) If the adjusted list contains a total of five or more responses, overall scores for the subject in question can meaningfully be calculated. Scores should then be adjusted to a basis of ten responses such that they become comparable between subjects.

Since the raw overall scores taken by themselves do not allow for a mean-ingful interpretation, we have worked out a standardization procedure. This procedure allows for a conversion of raw scores to C-scores, that is, standard-ized scores on an 11-point scale ranging from zero to ten, the mean score is five and standard deviation is two. Previous test results have made it possible to con-struct tables for score conversion for particular populations (for example, for candidate reserve officers).

Let us now consider some of the reliability and validity studies on our test method. We have run two *test-retest* reliability studies on candidate officers. The time lapse between testing and retesting was respectively between six and 12 months and two days. The former study yielded a test-retest correlation of .50, the latter of .77. We did a study on the *scoring reliability* by having five persons score the responses of 100 candidate professional officers. (Remember that in choosing synonyms the scorers have to make some subjective judgements.) If the poorest scorer was deleted from the correlation-matrix, all inter-scorer cor-relations per dimension were higher than .90. The *score reliability* was checked for by having two groups of ten judges score 24 randomly chosen expressions. The correlations were .95 for *extroversion*, .95 for *agreeableness*, .89 for *con-scientiousness*, .84 for *neuroticism*, and .85 for *general culture* (yielding .90 on average).

The *congruent validity* of the VZM was checked for in two studies. In a first study with 77 senior high school students we calculated correlations with the five-personality factor test (Elshout & Akkerman, 1975). Correlations per dimension were .71 for *extroversion*, .54 for *agreeableness*, .60 for *conscien-tiousness*, .55 for *neuroticism*, and .55 for *general culture*. In a second study (Böhrer & Van Den Broeck, 1986) with 63 candidate reserve officers, we calculated correlations between the eight VZM- dimensions and a set of ques-tionnaires in our selection procedure including the PMT of Hermans (1976) (measuring achievement motivation (P), and debilitating (F-) and facilitating (F+) fear of failure); the social anxiety scale of Willems, Tuender-De Haan, and Defares (1973); and the ABCA questionnaire of Böhrer (1980) (measuring social anxiety and self-confidence). The correlations are presented in Table 9.1. We hope these results can convince our readers that the VZM, as a new method of personality research, can stand up to questionnaires as far as reliability and validity goes. Aside from the theoretical justification of the VZM, which we have set up earlier in this chapter, some more practical considerations should also be mentioned in defense of our new test method: The VZM is not as time-consuming as questionnaires, as well as more attractive for the test takers, since they feel they can express themselves freely using their own words and there is no time pressure.

SKILLS: THEORY AND TEST DEVELOPMENT

Can Skills Be Reduced to Personality Features Plus Aspects of Intelligence?

I will now turn to the second theoretical issue. There are all kinds of folkpsychological features which we commonly ascribe to persons, for

Table 9.1

Correlations between VZM-Dimensions and Some Questionnaires

	VZM-Dimensions							
	SD	EXT	AGR	CON	NEU	GC	LEA	CRE
PMT P	.36	.24	.07	.32	-.17	.29	.34	-.02
F-	-.48	-.08	-.11	-.26	.41	-.33	-.41	-.04
F+	.21	-.01	.09	.21	-.14	.09	.23	-.03
Social Anxiety Scale	-.36	-.32	-.09	-.16	.32	-.32	-.45	-.25
ABCA Social Anxiety	-.42	-.58	-.30	-.24	.16	-.22	-.44	-.28
Self-Confidence	.37	.09	.01	.35	-.23	.25	.37	-.08

example, "Alice wants to become a medical doctor," "Joe felt lonesome yesterday night," "Mary is extremely bright at math," etc. Psychologists tend to cut up this set of features into two subsets, namely, a subset of cognitive features and a subset of noncognitive or personality features. Within the former set they distiguish between beliefs on the one hand, and intelligence on the other hand. The latter set contains a mixture of conative features, reflecting our values and aspirations, as well as *affective* features, depicting our emotional lives.

Now here is a problem concerning this dichotomy between cognitive versus noncognitive features. There exists a large set of mental ascriptions that can neither be classified as *pure* cognitive, nor as *pure* noncognitive features. I am particularly interested in the ascription of *skills*, say leadership skills, or the skill of learning a second language, etc. Now many such terms, which stand for skills, are *mixed* terms, that is, they neither denote *pure* cognitive, nor *pure* noncognitive features. Furthermore, for at least some subset of such mental ascriptions, it does not seem possible within folkpsychology to reduce such ascriptions nontrivially[4] to some mixed set of terms denoting either pure cognitive or pure noncognitive features. Many of our common terms denoting skills are irreducible in our folkpsychological language, that is, we simply lack the common vocabulary to replace names of skills by mixed sets of pure terms. In other words, for some names of skills, it is the case that the mental ascription "having such and such skill" (say, the skill for learning a second language) is *not* just reducible to a particular set of *common* terms denoting personality features (say "being sociable") and *common* terms denoting cognitive features (say, "having a good memory"). We want to say that having a skill for learning a second language is something more than just the sum of being sociable and having a good memory, and furthermore, whatever pure *common* terms we add to this sum, it will never equal the skill in question. In folkpsychology, there are at least some skills which are what they are and nothing else!

Now within proper psychology this state of affairs may prompt three kinds of responses:

1. We may decide to fine-tune our folkpsychological set of *pure* cognitive and (or) noncognitive vocabulary with the aim of making a reduction feasible within proper psychology. I propose to label this position *"narrow reductionism."*

2. We may decide to give up the reductionist project and include folkpsychological ascriptions of skills as irreducible terms in our psychological vocabulary. Let us call this the *antireductionist* position.

3. We may decide to construct a limited psychological set of special mixed terms, that is, terms that are in some sense elementary as well as general, though do include reference to *both* cognitive and noncognitive features. Such terms are said to denote types of *cognitive style.* The aim of constructing such set is to give the reductionist program more leeway: it is claimed that our folkpsychological ascriptions including pure cognitive terms, pure noncognitive terms, and (or) mixed terms of cognitive style. This position I propose to call *"broad reductionism."*[5]

Now, the assessment of (potential or actual) skills is a major concern in psychometric research, considering its practical applications for screening job applicants and for career-counseling. Now I believe that each of these theoretical stands on the issue of reductionism has its correlate in measurement techniques for the assessment of skills. In our own research we have developed two tests which respectively match the antireductionist and the broad-reductionist project. At first, however, I will relate the narrow-reductionist program to a classical technique in measurement theory, namely, empirical criterion keying.

The Narrow-Reductionist Project and Empirical Criterion Keying

Consider a (sympathetic) layperson's views on IQ-tests or personality tests in a context of selection or career-counseling. The layperson believes that the psychologist knows how to translate, say, "the skill of being a good sales manager" into a set of personality features, such as, "being meticulous" and "being assertive" and attributes of intelligence, "being proficient in mathematical reasoning." Furthermore the psychologist is thought to know how to translate raw test results into claims concerning the degree to which the testtaker can be assigned such personality features or attributes of intelligence. It is this *double-translation job* that makes it possible for the psychologist to uncover potential skills by means of apparently trivial test questions.

Now the layperson's views on measurement theory hinge on the belief (1) that skills can be translated into sets of common personality features and attributes of intelligence and (2) that sets of raw test results can be translated into claims concerning common personality features and attributes of intelligence. Counter to the first belief, I have already indicated that our folkpsychological vocabulary is too impoverished to allow for this sort of crude reductionism. Counter to the second belief, our concern about the content-validity of measurement techniques comes to show that this translation job is not always fully translucent.

Empirical criterion keying (cf. Anastasi, 1961, pp. 528-541 for empirical criterion keying on interest tests) cleverly avoids dependency on both precarious beliefs. How so? Some groups will score consistently higher on particular IQ questions and respond to personality tests in idiosyncratic ways. Groups of successful people within certain professions are presented with personality tests and/or IQ tests. For each such profession the psychologist then tries to identify particular scoring patterns. Within a context of selection, candidates for some jobs are then chosen because their particular scoring pattern most closely resembles the typical scoring pattern of some sample of persons successfully performing the job in question. Within a context of career-counseling, the future student's scoring pattern is compared to typical scoring patters for a number of jobs in order to locate the branch of study which would suit that student most.

The skill for performing certain professions is thus reduced to the fine-tuned language of scoring patterns in IQ tests and/or personality tests. In folkpsychology, it was impossible to express skills in terms of our common vocabulary denoting aspects of intelligence and personality features. Now empirical criterion keying (vis-à-vis skills) is the project in proper psychology which tries to reduce (potential) skills to a combination of aspects of intelligence and personality features, though described on terms of the sophisticated vocabulary of scoring patterns in psychological tests. In this project, the need for translating test results *or* skills in terms of folkpsychological terms denoting aspects of intelligence or personality features becomes obsolete.

It is clear that empirical criterion keying is an attractive project. There is the simplicity of analyzing skills only in terms of aspects of intelligence and personality features and the clever avoidance of the precarious belief in the possibility of a satisfactory execution of the double-translation job. But is this attractive project feasible? What price do we need to pay for such elegance? Here are a few conceptual considerations which made us decide not to choose for the narrow reductionist route in test development. First, there are certain personality features and/or aspects of intelligence which are causally, though not essentially related to the successful performance of a particular job. Personality features and aspects of intelligence which are essentially related to the successful performance of some jobs are—on the narrow-reductionist project—identical to the skill in question. Personality features and aspects of intelligence that are causally, though not essentially related to the successful performance of a job, are merely by-products of the skill in question. Empirical criterion keying does not allow us to distinguish between aspects of intelligence and personality features that are essentially versus nonessentially related to the successful performance of some job, or in other words, it cannot distinguish between the skill and its by-products. Please allow us to invoke a common and probably incorrect stereotype to illustrate this point. Librarians are often considered to be dull personalities. Now assume—counter to our honest expectations—that this personality feature would show up through empirical criterion keying. Now would we then say dullness is part of the skill of being a good librarian? I do not think so. Rather, we would say that the methodical work librarians do is casually efficacious in bringing about their dullness, but dullness is therefore not an essential feature of being a good librarian.

Second, it may well be the case that, for some jobs, the aspects of intellligence and personality features that are essentially related to "having the potential skill for some job" and "having the actual skill for some job" are very different or even incompatible. Consider some artistic discipline, for example, ballet. Most likely, creativity is an essential feature of the actual skill of a talented ballerina. But now this same creativity may be counterproductive in a beginning class for ballet. What we need the most in this context is an obedient and rigid performance of basic exercises. The personality feature that is essentially attached to the potential skill for ballet is thus very different from the personality feature attached to the actual skill for ballet. A talented ballerina does not start off as a creative dancer, but *acquires* this very creativity through rigidity. Now empirical criterion keying can solely tell us about the personality features and aspects of intelligence related to *actual skills*. For those personality features and aspects of intelligence for which it is essential that they are acquired *during* the job-training, empirical criterion keying can solely yield a misleading test procedure for assessing potential skills.

Finally, if skills *are* actually more than a combination of personality features and aspects of intelligence, empirical criterion keying on personality tests and IQ tests may leave some feature that is genuinely essential to having the skill in question, unassessed.

For all these reasons we have opted for steering our projects in test development away from the narrow-reductionist route.[6] We have worked out two tests that respectively match the antireductionist and the moderate-reductionist route. Both tests will now be discussed.

The Antireductionist Program and the ROMAT

In this section we will consider the methodological correlate of the antireductionist position. The antireductionist measurement theorist claims that any set of information obtained from cognitive tests and from personality tests (whether expressed in folkpsychological terms or in terms of scoring patterns) will be insufficient for properly assessing particular skills.

Consequently a very pragmatical path is opted for. The tests set up are miniature problem situations which the subjects are asked to solve either individually or collectively. This simulated problem situation is similar in relevant respects to *actual* problem situations which require for their solution the particular skill that is being tested for. Following this pragmatical route, we developed a particular test for the recruitment of candidate officers, namely, the ROMAT.

The ROMAT is a Group Situational Performance Test. A group of candidate officers, ranging from five to seven persons, are asked to assign sites in a building area on the outskirts of a city for a university, a sporting center, a hotel, etc. Each member in the group is responsible for one such institution. A set of instructions determines what features in a building site are desirable for each institution. There are some instructions all institutions have in common, for example, they all aim at being close to town. One instruction is specific for each institution. In this instruction it is stipulated that the institution in question should be as far as possible from some other institution, for example, the hotel should be as far away as possible from the cultural center, because of the noise problem from loud rock concerts. The subjects in our test are asked to take responsibility for one institution and only they read the instructions only for their own institution. After a few minutes they are asked to take place around a half-circular map of the building area. They are asked to start discussing the location of the institutions for which they are personally responsible as well as of three common projects: a parking lot, an industrial plant, and a storage place for inflammable chemicals. There are eight major and four minor building sites. (In Figure 9.1, B, C, D, and G represent the minor sites.) The institution for which the subjects are personally responsible can only be assigned only to major building sites, the common projects can be assigned to both major and minor sites. After 20 minutes the subjects should come up with some collective agreement.

There is both an objective and a subjective evaluation procedure for the ROMAT. Let us at first consider the subjective evaluation procedure. The test-leader fills out an observation sheet which contains a scale for (1) prominence, (2) efficiency, and (3) sociability, and which also leaves room for more specific remarks on each subject. These scales respectively indicate in how far each subject participates in the group discussion, comes up with good arguments in defense of his or her own interests, and comes up with good arguments in defense of collective interests.

Let us now consider the objective evaluation procedure. The test-leader makes a note of which institutions are ultimately assigned to which sites. Now, all instructions are phrased in terms of optimal distances, for example, the person who is assigned the university knows that this institution should be located as close as possible to town and as far as possible from the industrial plant. The distances between institutions which should be as far as possible from each other are added and subsequently, the distances between

Figure 9.1
The Map of the Building Site in the ROMAT

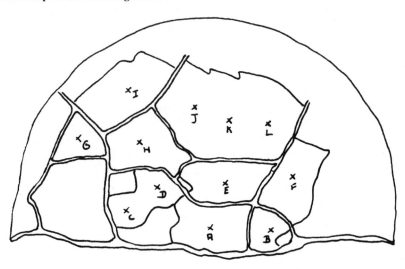

institutions which should be as close as possible to each other on each person's instructions, are subtracted from this sum. Some distances are measured as the crow flies (for example, the distance to an industrial plant), while other distances are measured via the connecting roads (for example, the distance to town). This procedure yields a score for each person indicating to what respect he or she has been successful in carrying through the instructions. All distances between sites are preprogrammed and scores are computer-calculated. So far no standardization procedure has yet been developed, so only intergroup comparisons between scores are genuinely meaningful.

The Broad-Reductionist Program and the RS1C

We will now turn to the methodological correlate of the broad-reductionist program. The broad reductionist claims that skills can indeed be reduced to more basic components, though this reduction should, aside from personality features and aspects of intelligence, also include modes of cognitive styles. Now the lesson to be learned from the broad-reductionist program for research on the measurement of skills is that our tests should not only focus on the former two variables, but should also have some interest in the latter variable.

We have taken the first steps in developing a computer-steered test which aims at assessing modes of cognitive style, at the same time as assessing aspects of intelligence. This test is essentially a letter substitution test and is named the RS1C. We will at first explain the mechanics of the RS1C and subsequently report some of the results from our pilot studies. The testtakers are placed in front of a personal computer. The PC explains to them that they will be presented with strings of six letters. Their task consists in reducing these strings to one particular letter by means of eight substitution rules Substitution rules are spelled out in the form XY = Z. A substitution rule, say PO = Q, applies to a particular string, if and only if, this string contains P and O at any place or in any order. Q is then inserted at the location of the first letter substituted for, whether this be P or O. Thus, given PO = Q, the letter string RSOQTP reduces to RSQQT. Subsequently the PC asks our subjects to type in their own names, copy eight strings of three letters and copy the eight substitution rules. Then they are shown how to solve one particular letter string. In the next step they will try to solve one very simple task. The computer teaches them how to make corrections, how to read the record of previously corrected substitutions on the screen, and how to ask for a new task if they decide to give up on the old one. After these instructions the actual test will start. There are six letter strings to be solved. There is no time pressure and the subjects are asked to try to solve all strings.

In a pilot study we checked for the correlations between mean solution time for each subject and results on IQ tests measuring verbal abilities, general reasoning, concentration, memory, and planning. This study yielded negative correlations ranging from − .23 to − .42. Surprisingly, the poorest correlation was with the concentration test, which was also a letter substitution test, though involving three-letter strings. Apparently letter substitution tasks, involving short versus long strings do not require the same skills for their solution.

Now we also searched for variables in our test which would correlate well with scores in a test measuring modes of cognitive style, namely, Herman's Achievement Motivation Test. Herman's scale expressing debilitating fear of failure correlated significantly with two variables in our own test. There was a significant positive correlation with the copying time of the three-letter strings (.22), of the substitution rules (.37), and with the standard deviation between solution times (.21). There was also a significant negative correlation between copying times for substitution rules and Herman's scale expressing facilitating fear of failure (− .22).

For some tasks the solution is more salient than for others in that (1) the order of the letters which should be substituted for in the successive steps on the solution route, is identical in the string and in the substitution rule; (2) these letters occur in the beginning of the string; and (3) these letters are located close to each other. Furthermore, for some task the solution route is the only route one can take on correctly applied solution rules, while for other tasks the solution route is only one alternative among the many branches that all lead to a dead end. We can safely assume that the difficulty of a substitution task is a negative function of the saliency of the correct solution route and a positive function of the complexity of the tree of branches. Now we are trying to construct an index of saliency and an index of complexity for a substitution task. Subsequently, we would like to investigate the following conjecture. While some persons may be more affected by increasing the complexity of the tree (considering their solution times) for equally salient trees, other persons may be more affected by increasing the nonsaliency (that is, decreasing the saliency) of the task (considering their solution times) for equally complex trees. Now, if this indeed would turn out to be correct, it would be interesting to investigate whether sensitivity to complexity or sensitivity to nonsaliency correlates with more familiar cognitive styles, for example, respectively with debilitating fear of failure and field-dependency. But these are only preliminary ideas for a new research project.

SUMMARY

In the first section, we discussed two situationist objections to personality theory namely that (1) all psychological explanation involving personality traits is merely ad hoc and (2) that there is no constancy to personality traits. Some personality theorists grant that these objections may hold for folkpsychology, but that they can be overcome if we carefully adjust the set of folkpsychological trait names for the purpose of doing proper psychology. We then discussed two such adjustment projects. Pervin (1976,1977) drastically expands the set of folkpsychological trait names by making them situation-specific. Tupes and Christal (1961) (and more recently McCrae and Costa, 1985a, 1985b, 1986, 1987) drastically reduce the set of folkpsychological trait names by isolating five factors in adjective checklists. We argued that the former project supersedes (at least partly) the charge of nonconstancy, though only at the cost of making the charge of ad hoc-ness become more acute. On the other hand, the latter project can (at least potentially) supersede the charge of ad hoc-ness, though only at the cost of making the charge of nonconstancy become more acute. I conclude that the set of folkpsychological trait names (or some close variant) may provide for an optimal balance between both situationist charges and may thus be the best conceptual candidate for psychological modeling. These theoretical ideas are then applied to the development of a free-format self-description method in personality testing.

In the second section we set out three stands in a debate on the ontology of skills: (1) skills are reducible to personality features and aspects of intelligence ("narrow reductionism"); (2) skills are not reducible at all, that is, skills are what they are and nothing less ("antireductionism"); and (3) skills are reducible to personality features, aspects of intelligence, and modes of cognitive style ("broad reductionism"). We went on to show that each ontological stand has its correlate in measurement theory. Then we discussed empirical criterion keying vis-à-vis the assessment of skills in reference to the narrow-reductionist project. The antireductionist stand is illustrated by means of the ROMAT, that is, a group situational performance test for leadership skills which was developed in the Section for

Psychological Research. As a correlate of the broad reductionist stand we presented some preliminary research on the RS1C, a computer-steered letter substitution test, which attempts to measure aspects of both intelligence and modes of cognitive style.

NOTES

1. We have decided not to discuss recent developments in questionaire methods in this chapter. For our own views on this topic, we refer to Claeys et al., 1981. Also worth mentioning is the excellent reader by Angleitner and Wiggins (1986), which covers a broad area in the current research on questionnaire methods.

2. Notice that folkpsychology is not synonymous with pop psychology, that is, the pseudo-psychology that we can find in popular magazines.

3. We have our doubts about the strength of this argument. It certainly is the case that a scoring pattern on the five-personality factors derived from personality features which hold across situations displays more constancy than a scoring pattern derived from highly situation-bound personality features. On the other hand, the problem of constancy still comes in for a person who comes out moderately *agreeable* since she is both highly *suspicious* as well as highly *attentive to people* across situations and consequently describes herself as such.

4. We have added this stipulation to rule out trivial reductions of the form "x is a person who is T (T being some mixed trait)" if and only if "x has the pure cognitive features of being T *and* x has the pure noncognitive features of being T."

5. A reductionist program faces the problem of isolating the components which make up skills. A related project is the study of the complex interrelationships between these components, that is, between personality features, aspects of intelligence and modes of cognitive style. This idea was presented by A. Heim, 1970, pp. 53-61. A. Böhrer and S. Van Den Broucke (1986) have done an extensive study on the correlations between a set of IQ-tests on the one hand and a set of personality tests on the other hand.

6. This conclusion may be slightly too hasty. Consider carefully the scope of the arguments against empirical criterion keying (ECK). The first argument does not only hold against ECK on a narrow reductionistic route, but also against ECK on a broad reductionistic route. Indeed the problem of distinguishing between essential and nonessential features in a scoring pattern also comes in if we bring in variables of cognitive style. The second argument against ECK is actually a more general warning against equating potential skills and actual skills in career-counseling or screening job applicants. Claim (c) against ECK strictly focuses on ECK on a narrow reductionistic route. This claim solely provides for an argument against narrow reductionism if it is indeed the case that ECK for skills on IQ-tests and personality-tests yields relatively poor validity-coefficients. We thus do not intend to take a definite stand in the debate concerning the analysis of skills, since the first two arguments are not strictly arguments against narrow reductionism, while the final claims indeed targets narrow reductionism though is in need of empirical support.

REFERENCES

Allen, B. M., and Potkay, C. R. (1973). Variability of Self-Description on a Day-to-Day Basis: Longitudinal Use of the Adjective Generation Technique. *Journal of Personality, 41*, 638-652.

_____. (1977). The relationship between AGT Self-Description and Significant Life Events: A Longitudinal Study. *Journal of Personality*, *45*, 207–219.

Anastasi, A. (1961). *Psychological Testing* (2nd ed.). New York: Macmillan.

Angleitner, A., and Wiggins, J.S. (eds.). (1986). *Personality Assessment via Questionnaires—Current Issues in Theory and Measurement*. Berlin: Springer Verlag.

Böhrer, A. (1980). Social Anxiety, Self-Confidence and Military Leadership. *Proceedings of the 22nd Annual Conference of the Military Testing Association, 1*, 91–102.

Böhrer, A., and Van Den Broucke, S. (1986). Personality and Test Performance. In S.E. Newstead, S.H. Irvine, and P.L. Dann (eds.), *Human Assessment: Cognition and Motivation*. Dordrecht, Martinus Nijhoff, 287–298.

Claeys, W., De Boeck, P., and Böhrer, A. (1981), How to Improve the Validity of Personality Questionnaires. *Proceedings of the 23rd Annual Conference of the Military Testing Association, vol. 1*. Coordinated by the U.S. Army Research Institute for the Behavioral and Social Sciences, Arlington, Virginia.

Claeys, W., De Boeck, P., Van Den Bosch, W., Biesmans, R., and Böhrer, A. (1985). A Comparison of One Free-Format and Two Fixed-Format Self-Report Personality Assessment Methods. *Journal of Personality and Social Psychology*, *49*(4), 128–139.

Elshout, J. J., and Akkerman, A. E. (1975). *Vijf-Persoonlijkheidsfactoren Test (SPFT) Handleiding*. Nijmegen: Berkhout.

Heim, A. (1970). *Intelligence and Personality—Their Assessment and Relationship*. Harmondsworth: Penguin.

Hermans, H. J. H. (1976). *P.M.T. Prestatiemotivatietest—Handleiding*. Amsterdam: Swets & Zeitlinger.

Le/De Romat (1986). Manuel pour les observateurs/Handleiding voor Observatoren. Section for Psychological Research (Belgian Army), mimeo.

McCrae, R. R., and Costa, P. T. Jr. (1985a). Comparison of the EPI and Psychoticism Scales with Measures of the Five-Factor Model of Personality. *Personality and Individual Differences*, *6*, 587–597.

_____. (1985b). Updating Norman's "Adequate Taxonomy": Intelligence and Personality Dimensions in Natural Language and Questionnaires. *Journal of Personality and Social Psychology*, *49*(3), 710–721.

_____. (1987). Validation of the Five Factor Model of Personality Across Instruments and Observers. *Journal of Personality and Social Psychology*, *52*(1), 81–90.

McCrae, R.R., Costa, P.T. Jr., and Busch, C.M. (1986). Evaluating Comprehensiveness in Personality Systems: The California Q-Set and the Five Factor Model. *Journal of Personality*, *54*(2), 430–446.

Molière, J.B. (1947 [1673]). *Le malade imaginaire*. Bruxelles: Les Editions de la Cité Genève.

Pervin, L.A. (1976). A Free-Response Description Approach to the Analysis of Person-Situation Interaction. *Journal of Personality and Social Psychology*, *34*(3), 465–474.

_____. (1977). The Representative Design of Person-Situation Research. In D. Magnusson and N. Endler (Eds.), *Personality at the Crossroads: Current Issues in Interactional Psychology*. New York: Wiley, 371–384.

Potkay, C. R., and Allen, B. P. (1973). The Adjective Generation Technique: An Alternative to Check Lists. *Psychological Reports, 32,* 457–458.

Tupes, E. C., and Christal, R. E. (1961). *Recurrent Personality Factors Based on Trait Ratings.* USAF ASD Technical Reports (No. 61–97).

Vrije Zelfbeschrijvingsmethode. (1987). Lisse (Netherlands): Swets Test-Services. In translation, *Free-Format Self-Description Method.* (1987). Kortenberg (Belgium): Center for Basic Interactive Research. *L'auto-description libre.* (1987). Kortenberg (Belgium): Center for Basic Interactive Research.

Willems, L. F. M., Tuender-de Haan, H. A. and Defares, P. B. (1973). Een Schaal om Sociale Angst te Meten. *Nederlands Tijdschrift voor Psychologie, 28,* 415–422.

10   Dimensions of Job Performance

Kevin R. Murphy

Preparation of this chapter was partially supported by an ASEE/Navy Summer Faculty Fellowship.

INTRODUCTION

Job performance is a familiar concept—one so familiar that there is little apparent need to provide a formal or complete definition of precisely what is meant by the term "performance." In this chapter, I will argue that there *is* a need for such a definition, that a satisfactory definition does not presently exist, and that the methods which have been used in attempting to define job performance are not sufficient for the task. A framework is proposed that defines the domain of job performance by specifying the major dimemsions of performance as well as the relations among those dimensions.

Is a Definition Useful?

Most of the applications of psychology in work settings are carried out in an effort to maximize job performance. These applications revolve around four basic questions: (1) How should we measure performance? (2) How can we select the applicants with the highest potential to perform well? (3) How can we train workers to increase their level of performance? and (4) What are the likely consequences of systematic efforts to increase performance? It is difficult to envision a sensible answer to any of these four questions in the absence of an acceptable definition of performance.

The need for a good definition of the performance domain is most obvious when we consider the extensive body of research dealing with the measurement of job performance. Attempts to determine the validity of specific performance measures have been infrequent, and have been characterized by serious statistical and methodological shortcomings (Landy & Farr, 1983; Saal, Downey, & Lahey, 1980). Some methodological improvements have been suggested (Borman, 1977; Murphy, Garcia, Kerkar, Martin, & Balzer, 1982), but these are limited in their applicability. A construct validation strategy has been suggested (James, 1973; Smith, 1976), but has rarely been applied with any success. It has been possible to apply some of the *methods* used in construct validation (for example, multitrait-multimethod analyses; see Lawler, 1967, Kavanaugh, MacKinney, and Wolins, 1971), but the construct validation strategy cannot work in the absence of a well-specified definition of the construct.

Consider the problem of determining whether supervisory performance ratings provide a valid measure of job performance. It is difficult to determine whether a specific measure or set of measures provides a valid indication of a person's performance in the absence of a definition of the performance domain. In particular, there is generally no basis for determining whether: (1) any given measure is within or outside of the domain of performance, (2) a given set of measures spans the major dimensions of that domain, or (3) the relations among criterion measures or between predictors and specific criteria are consistent with the definition of the domain of performance. This last point is illustrated graphically in the debate over the proper measurement and treatment of halo error in performance ratings (Harvey, 1982; Landy, Vance, Barnes-Farrell, & Steele, 1980; Landy, Vance, & Barnes-Farrell, 1982; Murphy, 1982; Saal, Downey, & Lahey, 1980). Halo measures have almost universally depended on arbitrary statistical assumptions, largely because of the fact that no theoretical basis exists for determining how much true halo (Cooper, 1981) ratings *should* show. Measures of other "rater errors," such as leniency and central tendency, show the same problems, for largely the same reasons (Murphy & Balzer, 1981; Saal, Downey, & Lahey, 1980).

The principal goal of personnel selection is to select individuals who are most likely to perform well. It is well known that tests of cognitive ability provide a valid basis for selection decisions (Hunter & Hunter, 1984). However, our understanding of the links between ability and performance have hardly progressed beyond the simple principle that workers who receive high test scores also tend to receive high performance ratings. Dunnette (1976) has noted a marked asymmetry in our understanding of the ability and performance domains. Although many points of controversy remain, the broad outlines of the content and structure of the domains of mental ability, physical strength, and psychomotor performance are well understood. Much less is known about the domain of job performance. Thus, it is difficult to construct or test sophisticated theories linking specific or multiple abilities, skills, and personal characteristics with performance. Our lack of well-articulated definition of the performance domain also hinders efforts to design effective training programs. Many training courses are aimed at the acquisition of highly specifc skills, which may not be relevant to most aspects of performance. In the absence of a definition of the performance domain, there may be no way of determining which of the more general skills, (for example, written communication, supervision) are most relevant to the job, or which areas should be targeted for training.

A substantial literature on the utility of selection tests, and of various productivity improvement programs (for example, programs based on incentive pay) suggests that both tests and productivity improvement programs lead to substantial increases in performance (Guzzo, Jette, & Katzell, 1985; Hunter & Hunter, 1984). The financial impact of improvements in performance has been widely researched (Hunter & Schmidt, 1982; Murphy, 1986; Schmidt, Hunter, McKenzie, & Muldrow, 1979), but there has been little attention given to the nonfinancial consequences of improved performance. It seems likely that an organization made up of good performers would be both quantitatively and qualitatively different from an organization made up of poor performers (Mitchell & Kalb, 1982; Mitchell & Wood, 1980). The difference would not simply be in the number of things produced, but would also be manifest in a wide range of interpersonal processes in the organization. For example, the supervisor's job would be quite different if all poor performers were replaced by good performers. In the absence of a reasonably complete definition of performance, it is difficult to describe in any concrete way the full range of consequences which might be associated with substantial changes in performance.

Does a Definition Exist?

There is extensive literature dealing with task performance, job performance, and productivity. A problem common to all three of these literatures is a lack of agreement concerning the variables which underly and define the meaning of "performance." This is most obvious in research on productivity. Three strategies seem to dominate in defining productivity: (1) to leave productivity undefined (Sutermeister, 1976), (2) to admit that the concept is ill-defined and propose a vague definition (Muckler, 1987), or (3) to propose several definitions, each of which applies in different circumstances (Guzzo, Jette, & Katzell, 1985). Thus, it is difficult to agree whether productivity is going down (and if so, by how much), whether productivity enhancement programs really work, and most important, what the consequences would be if productivity levels were to change substantially.

In research on task performance, definitional problems are not so readily apparent. Yet, as the work of Fleishman and his associates has illustrated, the classification and definition of task performance is exceedingly complex (Fleishman & Quaintance, 1984). Part of the problem is that it is difficult to define the boundaries of complex tasks. That is, it is often difficult to delimit task behavior from nontask behavior; this problem becomes especially acute when task behavior includes mental as well as physical labor. In addition, performance on several common tasks is both complex and multidimensional, especially when considered in a real-time context in which several tasks must be attended to. Thus, it can be difficult to determine precise links between an individual's behavior and his or her progress in completing a specific task.

There is an extensive literature dealing with the prediction and measurement of job performance. For example, Landy and his colleagues have reviewed research on performance rating and have suggested several innovative directions for further research and application (Landy & Farr, 1980, 1983; Landy, Zedeck, & Cleveland, 1983). In the area of prediction, Schmidt, Hunter, and their colleagues have reviewed hundreds of studies which suggest that ability tests provide valid predictions of job performance (Hunter & Hunter, 1984; Pearlman, Schmidt, & Hunter, 1980; Schmidt & Hunter, 1977, 1981). Both of these literatures are notable for the fact that performance, which is the central concern of the research reviewed, is rarely, if ever, defined. As a result, we know a great deal about predicting future performance, and also know a great deal about measuring past performance, but know little about what we have predicted or measured. Campbell (1983) notes that we have been quite successful in modeling and defining different parts or aspects of performance, but that little effort has gone into defining the overall performance domain.

Job Performance versus Task Performance

The preceding section suggests that little attention has been directed to defining the domain of job performance. One possible explanation for this is that job performance is typically equated with task performance. That is, it is often assumed that job performance can be adequately defined in terms of the job incumbent's success in carrying out the tasks which are included in a formal job description. From this perspective, the growing body of research dealing with the determinants and the nature of task performance (Fleishman & Quaintance, 1984; Salvendy & Seymour, 1973) provides the key to understanding job performance. If job performance is equated with task performance, our research priorities in this field should be directed at developing better job descriptions and at aggregating measures of performance on several separate tasks into an overall measure of job performance.

There are several reasons to believe that job performance *cannot* be equated with task performance. First, most observations of work behavior confirm the common perception that workers spend relatively little time performing what would be regarded as tasks. For example Bialek, Zapf, and McGuire (1977) reported that enlisted infantrymen spent less than half of their work time performing the technical tasks for which they had been trained; in many cases, only a small proportion of an enlisted person's time was in any way devoted to accomplishing the tasks specified in his or her job description. Campbell, Dunnette, Lawler, and Weick (1970) noted similar patterns for managers. A substantial part of the manager's day is spent doing things that cannot be unambiguously linked to accomplishment of specific tasks. The fact that most people's work time is not devoted solely to tasks has serious implications for several aspects of criterion development. For example, many methods of job analysis are based explicitly on the assumption that most, if not all of a worker's time is spent working on identifiable tasks (for example, the Air Force's task analysis system, see Christal, 1974). Since much of the workday is spent doing something outside of the typical domain of tasks, indexes of the percentage of time spent on different tasks present a warped view of the activities actually carried out by workers. Unless you are willing to ignore much of what a person does at work, it is difficult to equate job performance with task performance.

Second, many of the performance evaluation systems currently in use include specific measures or indexes which are only tangentially related to task performance. Examples include measures of absenteeism and turnover, as well as supervisory ratings of broad traits such as dependability or motivation. Admittedly, the use of such measures does not prove that the job performance domain is broader than the task performance domain; it is possible that these measures, although widely used, are invalid. Nevertheless, the widespread use of measures which do not relate directly or solely to the accomplishment of tasks does suggest that the job performance domain is *perceived* to be broader than the domain of task performance.

A third argument for assuming that job performance cannot be defined solely in terms of task performance is that job performance must be defined over longer time periods and in relation to more organizational units than is true for task performance. This can be seen most clearly by considering the case of a plant manager who successfully meets a first-quarter production quota by depleting all available reserves of material and by diverting resources from other parts of the organization. Although the manager has successfully achieved the main task, the long-term implications of such task performance is clearly not favorable. In many jobs, it would not be difficult to provide examples of individuals who successfully completed most tasks, but whose performance was judged to be low.

PERFORMANCE DIMENSIONS AND PERFORMANCE CONSTRUCTS

In defining the domain of job performance, two questions must be answered. First, what are the boundaries of this domain? In other words, what is the "stuff" of job performance? Second, how is the domain organized? Both of these questions have been addressed in part by several researchers.

What Constitutes Job Performance?

Smith (1976) has noted that performance could be defined in terms of behavior, in terms of results of behavior, or in terms of an individual's contribution to organizational effectiveness. Kavanaugh (1971) has argued that the performance domain should also include work-related traits. In general, the question of whether performance should be defined in terms of traits, behaviors, or the outcomes of behaviors varies as a function of the job in question. In some jobs, such as those held by individual craft workers, there is a close connection between job behaviors and results. In other jobs, such as those on assembly lines, results (productivity) are controlled largely by situational rather than individual factors. In some jobs, performance is defined in terms of what people do, while in others (for example, fire fighters, members of the military) performance is defined partially in terms of readiness rather than in terms of specific job tasks.

The fundamental question in determining the content of the job performance domain is whether performance should be defined in terms of behavior or in terms of the results of behavior (James, 1973; Smith, 1976). From the organization's point of view, there is a strong temptation to define performance in terms of results—most managers like to think of themselves as members of a hard-driving, results-oriented organization. There are, however, several reasons for defining performance in terms of behavior rather than in terms of results. First, an exclusive emphasis on results is likely to lead to behaviors that are dysfunctional for the organization. Landy and Farr (1983) note that if performance is defined exclusively in terms of countable outcomes (that is, results), job incumbents will be strongly motivated to maximize those outcomes at the expense of other activities (for example, maintenance, planning, conservation) which are vital to the organization. Second, results are more complexly determined than behaviors, in that results are a joint function of what the person does and the situation in which one does it. Thus, until we more fully understand the domain of job behaviors, it may not be possible to fully understand the joint effects of behaviors and situations. Third, as psychologists, we have more to contribute to the understanding of behaviors than to the understanding of behavioral outcomes. The methods and theories of psychology are more useful when applied to a domain of job behaviors than to a domain of outcomes.

The consideration outlined above suggests that job performance should be defined as a domain of behaviors which occur on the job, or in conjunction with the job. As will be noted in a later section, "job behavior" can be broadly defined to include highly specific acts as well as general patterns of behavior (that is, traits). Several types of behaviors might be included in this domain; research aimed at determining the dimensions of criteria is relevant in defining the boundaries of that behavioral domain.

Research on the dimensions which underlie various measures of job performance has lead to two general conclusions. First, it is widely agreed that performance is multidimensional (Bass, 1982; Dunnette, 1963; Pickle & Friedlander, 1967; Ronon & Prien, 1966; Seashore, 1975). Second, there appears to be no strong general factor underlying analyses of the most common performance measures (James, 1973; Smith, 1976). Taken together, these two conclusions might lead to the further conclusion that concepts such as "overall performance" or "general level of performance" are potentially meaningless. However, these concepts, widely used in practice (Landy & Farr, 1983), are thought to be necessary for administrative purposes (Schmidt & Caplan, 1971), and underlie the vast body of literature tying personnel and selection to outcomes such as productivity (Hunter & Hunter, 1984; Hunter & Schmidt, 1982; Murphy, 1986; Schmidt & Hunter, 1977, 1981; Schmidt, Hunter, McKenzie, & Muldrow, 1979). In part, the gap between the body or research that suggests that there may be no such thing as "overall performance" and the body of research that treats "overall performance" as its central concern can be traced to differences in method and focus. As the section that follows will suggest, the failure to find a strong general factor may say more about the shortcomings of the methods used to define the performance domain than about the domain itself.

Methods of Determining Performance Dimensions

Four methods could be applied in determining the behavioral dimensions that define the domain of job performance. First, you could obtain a representative sample of the different measures that are commonly used in measuring job performance and apply factor analysis or some related technique to the intercorrelations among these measures. Rush (1953) applied this strategy in studying sales criteria, and suggested that the underlying dimensions of sales success could be defined as: (1) Objective Achievement, (2) Learning Aptitude, (3) General Reputation, and (4) Sales Technique. Ronan (1963), Richards, Taylor, Price, and Jacobsen (1965), and Turner (1960) reported similar analyses, yielding from four to 29 factors, depending on the job category and the criterion measures contained in each study.

The factor analytic approach is basically inductive, in that it attempts to discover the nature of performance by analyzing various measures of performance. The difficulty with this approach is that it involves two untestable, and probably untenable, assumptions. First, one must assume some level of content or construct validity for each of the measures included in the analysis. If these measures will be used to define the dimensions of performance, it must be assumed that they all have something to do with performance. More important, this method assumes that the set of performance measures studied spans the domain of performance; performance dimensions such as work avoidance will emerge if absenteeism measures are included, but will not emerge if data from personnel files are not included in the analysis. Thus, the decision to either include or exclude a specific measure from your study (for example, a particular measure of absenteeism) represents an a priori decision about the boundaries and structure of the performance domain. If you knew which measures were and were not included in the content domain, and also knew which measures must be included to obtain a representative sample of that domain, it is unlikely that you would need to do a factor analysis in order to determine the nature and structure of the domain.

A very different method of determining the dimensions of job performance is to observe and analyze worker's behavior, in other words, to conduct some sort of job analysis. For example, Hemphill (1959, 1960) suggested that performance as an executive involved dimensions such as long-range planning, exercising power, and community involvement. McCormick and his colleagues have developed structured job analysis questionnaires which assess the behavioral elements common to performance in a wide variety of jobs (McCormick, 1979; McCormick, Jeanneret, & Mecham, 1972). Factor analyses of these questionnaires would certainly provide useful information regarding the dimensions of job performance. The problem with the job analytic approach is that it is difficult to obtain meaningful observations without providing some structure, and that the structure of the measurement instrument ifself will have a strong impact on your eventual conclusions regarding the dimensions of job performance. For example, questionnaires used to analyze Air Force jobs concentrate exclusively on discrete tasks (Christal, 1974). The dimensions that emerge from analyses of these questionnaires will therefore be task dimensions. The PAQ samples a broader array of behaviors, but concentrates heavily on issues such as the use of tools and machines. The content of the PAQ questions will affect the pattern of results coming from an analysis of PAQ responses, In a general sense, the job analyst's decision of what to observe and how to quantify those observations will have some impact, and the dimensions uncovered through job analytic methods cannot be regarded as simply dimensions of job performance. Rather, they represent an interaction between the structure of the domain and the structure of the instrument chosen to analyze the domain. As was true with factor analyses of performance measures, dimensions based on job analysis data may say more about the measures analyzed (and not analyzed) than about the domain. Unless an adequate sampling of the domain can be guaranteed prior to the analysis, these methods may be insufficient for revealing the dimensions of job performance.

The methods of human factors engineering present another approach to defining the dimensions of job performance (Chapanis, 1976; Campbell, 1983). Rather than using statistical analyses of performance measures or observations to uncover these dimensions, it might be possible to analyze the physical and cognitive and psychomotor demands of tasks, and to derive dimensions on the basis of task demands (Fleishman & Quaintance, 1984). There are, however, two limitations to this approach. First, this approach works well with simple tasks, but is difficult to apply to complex, poorly defined tasks (Campbell, 1983). Second, this approach limits its focus to the domain of task performance. It would be difficult to apply this approach to work behaviors as absenteeism, that are believed to be important aspects of performance.

A fourth approach to defining performance dimensions is the construct-oriented strategy. Although this approach has been widely advocated (James, 1973; Smith, 1976), it has rarely been applied. Here the task of defining the dimensions of job performance could be regarded as an exercise in construct explication (Nunnally, 1978). Using this approach, performance dimensions are defined rather than discovered; the set of dimensions decided upon depends on the definition employed for the construct ''performance.''

The construct-oriented model is appealing for several reasons. First, a construct-oriented strategy has long been recognized as the optimum method for assessing the validity of performance *measures* (Guion, 1980; James, 1973). The same model which is so highly recommended for the specific problem of assessing individual measures is obviously relevant to the more general problem of defining performance dimensions. Second, as noted above, the construct-oriented approach provides a way of avoiding the trap implicit in the inductive approach—that performance dimensions can be discovered, if only the right methods are applied. The construct-oriented approach demands that these dimensions be defined as part of the process of construct explication. Furthermore, the construct-oriented approach does not place arbitrary statistical limitations of the nature of the performance dimensions. Factor analysis produces dimensions that are orthogonal, and which account, in descending order, for the greatest proportion of the common variance, etc. While these statistical features are useful, there is rarely any reason to suspect that the underlying dimensions of performance, or of other behavioral constructs, *do* show any of these characteristics.[1] Dimensions that are defined rather than derived need not reflect these potentially arbitrary statistical limitations. Finally, the construct-oriented strategy makes explicit the assumption that performance *is* a construct, and that the ultimate definition of performance dimensions depends entirely on our conceptual definition of performance. In the section that follows, a construct-oriented strategy is applied to define performance, to define the dimensions of performance, and to specify the interrelations among the dimensions of job performance.

DEFINING THE DIMENSIONS OF JOB PERFORMANCE

Rather than using an inductive approach to "discover" the dimensions of job performance, one might define those dimensions in terms of their logical relations to the construct of job performance. The preceding section suggests that these dimensions should be behavioral dimensions, but says little about what behaviors should or should not be included in a definition of performance. Astin's (1964) definition of conceptual criteria provides a useful tool for defining the behavioral elements of job performance.

Astin (1964) noted that in developing criteria, we must identify the relevant goals of the sponsor or of the measure. In the context of work, the relevant goals of the organization would include both short-term goals, such as the successful completion of specific tasks, as well as long-term goals, such as the maintenance of effective relations between work group, departments, etc. According to Astin (1964), the conceptual criterion is nothing more than a verbal abstraction of the relevant goals or the outcomes desired; the set of possible criterion measures would include any observable index or state that is judged relevant to the conceptual criterion. If we substitute "performance construct" for "conceptual criterion," and "performance dimensions" for "criterion measures," it becomes possible to define the domain of performance and to indicate the set of behaviors that are included in that domain. The performance domain is defined here as the *set of behaviors that are relevant to the goals of the organization or the organizational unit in which a person works*.

In order to specify the range of behaviors which define performance, one would have to know the relevant goals of the organization. Note here that the global set of goals which define the overall effectiveness of the organization (for example, maximize after-tax profits) are not as relevant as the set of goals that are defined for an incumbent in a specific position within the organization. That is, the organization defines a set of goals to be met by the incumbent in each job, and the relevant goals may vary considerably from job to job, or across different levels in the organization. This is particularly true when specific task goals are considered, but will also be true for several nontask goals. For example, one goal that is likely to be broadly relevant, but which is not tied to any specific task, is that incumbents must maintain effective interpersonal relations with their coworkers and with other organizational members with whom they interact. The type and extent of these interpersonal contacts will vary across departments and across levels of management. One might infer that the skills and behaviors which contribute to successful maintenance of interpersonal relations will vary from job to job. Nevertheless, within the great majority of jobs, this general class of behaviors is likely to represent one aspect of effective performance.

Since the relevant goals of organizations and of organizational units differ, it may not be possible to draw up a *completely* general definition of the dimensions of job performance. Nevertheless, there are enough broad similarities in organizational goals and in job demands to justify a framework that is relevant for defining performance in a large class of jobs. For example, the performance dimensions that are relevant for describing skilled craft jobs (for example, electrician, plumber) would vary in some of their specifics, but would probably show considerable communality.

To illustrate the application of a construct-oriented approach to defining the dimensions of performance in a large class of jobs, a framework was developed for describing the domain of performance as an enlisted person in the Navy (Murphy, 1985). This category of jobs includes administrative, technical, and warfare-oriented jobs on both ship and shore; job titles range from Boatswains Mate to Data Processing Technician, from Missile Technician to Aviation Storekeeper. Although the specific goals associated with individual jobs vary, there are some general goals that are relevant to almost all enlisted positions. One such goal is *readiness*. In peacetime, one of the major goals of the Navy is to maintain a high state of readiness, of both machines and personnel. This is one of the reasons that the services place a heavier emphasis on training and practice than is true in industry. A second relevant goal is to attain and maintain *technical proficiency*. Since a large number of enlisted personnel serve for periods of four to six years, it is important that recruits quickly attain the complex technical skills needed to perform many of the jobs on both ship and shore. A third relevant goal is *effective teamwork*. In highly interdependent systems, such as ships or aviation wings, it is natural to emphasize the performance of units rather than the performance of individuals; an individual who perform his or her task effectively, but who impedes the work of the unit as a whole is likely to be viewed as an ineffective worker. Finally, as is the case in almost any job, *successful accomplishment of major job tasks* is a relevant goal.

Having defined a set of relevant goals, it is possible to define the set of behaviors that are necessary to meet these goals. First, enlisted personnel must acquire the general skills and specific knowledge which are needed to accomplish the tasks that are associated with their job. These skills and knowledge must then be effectively combined to achieve proficiency in carrying out these tasks. There is a useful distinction between what a person can do and what a person actually does on the job. For this reason, it is useful to distinguish proficiency from the actual performance on the job. This distinction will be discussed in more detail below.

There is a set of nontask behaviors that is clearly relevant to the goals stated above. For example, job incumbents must maintain interpersonal relations in the work setting, at least to the degree that the performance of jobs or tasks involves communicating and cooperating with others. In some settings, interpersonal relations may be confined to exchanging job-related information with superiors or with coworkers. In other settings, in which the job entails some supervisory or managerial functions, the range of relevant interpersonal behaviors might be significantly larger. In addition, there are behaviors that affect the likelihood that tasks will be successfully accomplished, but which are not in themselves task-oriented. First, there is a cluster of work-avoidance behaviors such as absenteeism. The term *down-time behaviors* is used here to describe behaviors that result in the incumbent being absent from the work site, or which result in seriously impaired levels of performance. This group of behaviors would include alcohol and drug abuse, as well as behaviors leading to court martials or to disciplinary confinement. Second, there is a cluster of behaviors that leads to a clear risk of productivity losses, damage, or other setbacks. These destructive/hazardous behaviors would include safety and security violations, accidents, and willful destruction or tampering with equipment, materials, or job information.

This total set of behaviors, when combined, lead to two distinct end states. First, these behaviors define a person's *Overall Effectiveness* in his or her position. In contrast to the conclusions implied by the factor analytic research reviewed earlier, the present framework will suggest that overall performance is a meaningful concept, and that the general performance level is affected by several distinct classes of behavior. In particular, Overall Effectiveness is a complex function of: (a) task performance, (b) destructive/hazardous behaviors, (c) down-time behaviors, and (d) interpersonal relations.

Behaviors in categories b, c, and d are likely to have a different functional relationship to job performance than are behaviors in category a. In particular, destructive/hazardous and down-time behaviors serve to limit performance if they are present. The absence of these behaviors, however, does not guarantee good performance. The functional relationship between interpersonal relations and performance will vary according to the job context. In some cases, poor interpersonal relations will hinder job performance, but better-than-average interpersonal relations will not lead to increased performance. Task performance, on the other hand, is likely to have a consistent positive relationship with job performance.

The second relevant end state is a person's *Overall Value* to the organization (in this case, to the Navy). Overall value is a joint function of what one does and how well one does it. Some positions will be viewed as more critical than others, because of the tasks that are performed, the skills or training which are required, or the difficulty in replacing individuals in that position. A person's overall value is therefore not strictly part of the performance domain, but rather reprensents an evaluation of the organizational consequences of effective performance. Overall Effectiveness, in contrast, represents a summary of the total set of behaviors which comprise the performance domain. That is, overall effectiveness represents the degree to which the job incumbent satisfies the relevant goals associated with his or her job.

The relations among the different classes of behaviors that comprise the performance domain are illustrated in Figure 10.1. Several aspects of these relationships should be noted. First, a definite causal hierarchy can be

Figure 10.1
Dimensions of Performance for Navy Enlisted Ranks

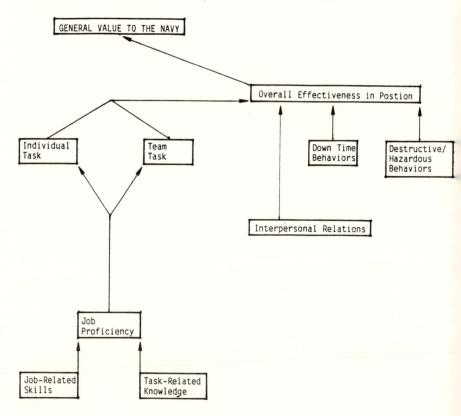

established among these behaviors. In particular, task performance depends on interpersonal relations and job proficiency. Job proficiency, in turn, depends of the acquisition of both general skills and task-related knowledge. Second, the causal flow is unidirectional, and can be broken down into two main paths. First, there is a set of nontask behaviors that independently affect overall effectiveness. Second, there is a set of behaviors that, taken together, define task performance, which in turn affects overall effectiveness. The contribution between each of these behaviors and effectiveness are described below.

Behaviors that Limit Performance

There are several types of behaviors that serve to limit job performance, even when major tasks are carried out effectively. Down-time and destructive/hazadous behaviors belong in this group. This class of behaviors is unique in terms of its functional relationship with other types of criteria. These behaviors are also (hopefully) unique in that they have low base rates. One implication is that correlational methods will do a poor job summarizing the effect of these behaviors and the place of this class of behaviors in the total performance domain.

Interpersonal Relations

This includes effective communications as well as effectiveness in dealing with others. Interpersonal relations contribute to effective team performance, but may have only a limiting effect on other performance categories. That is, effective interpersonal relations contribute to team task performance, but do not contribute in any direct way to overall effectiveness. Poor interpersonal relations, on the other hand, have a direct negative effect on both team task performance and overall effectiveness.

TASK BEHAVIORS

Individual and Team Task Performance

The actual accomplishment of specific tasks in everyday work settings is a function of several factors. First, individuals must be able to perform the task. Second, they must be motivated to perform the task. Third, they must have the resources available necessary to perform the task. Finally, they must correctly perceive which tasks need to be done. Task performance refers to what individuals *actually* do; job proficiency refers to what they *can* do.

Job Proficiency

Both task performance and job proficiency criteria are concerned with the same subset of behaviors (task behaviors). The difference between the two lies in the conditions under which these tasks are performed. Task performance criteria deal with the performance under normal working conditions. Job proficiency criteria deal with task performance under controlled conditions in which: (1) there are no competing task demands, (2) the subject is motivated to perform at the maximum level and to follow standard procedures, and (3) performance is continually monitored and evaluated in considerable detail.

Job-Related Skills and Task Knowledge

In order to attain proficiency, a body of factual and procedural knowledge, together with a set of skills ranging from psychomotor skills to analytic skills, must be mastered. These skills and knowledge bases represent basic components; proficiency is achieved only when these basics are effectively mastered and combined. The principal distinction between skills and knowledge is that the former necessarily involve performance of some activity, and are likely to have some generality, whereas the latter involves command of facts and details which are tied to specific tasks.

ORGANIZATION OF THE FRAMEWORK

Three distinct principles can be used to organize and describe the relationships among work behaviors shown in Figure 10.1. First, as shown in Figure 10.1, they can be hierarchically organized in such a way that each behavior is causally related to other behaviors or end sites, and in which causation always flows upward. Second, as shown in Figure 10.2, behavior can be classified according to sites; some behaviors occur only at the actual work site, whereas others can be assessed off-site. Third, as shown in Figure 10.3, behaviors can be classified as either task-related or as behaviors that do not directly relate to task performance, but which are nevertheless relevant in defining performance levels. Finally, as shown in Figure 10.4, behaviors can be cross-classified according to both site and task nontask designations. Each of these classifications of the performance domain is potentially useful for assessing the adequacy of specific measures of performance, for developing theories of performance, and for evaluation of programs designed to increase productivity or performance.

PERFORMANCE DIMENSIONS AND
PERFORMANCE MEASURES

Several types of performance measures have been used in organizations and in the military, ranging from job knowledge tests to global evaluations provided by supervisors. These measures would not necessarily provide equivalent information, nor would they provide equivalent coverage of the performance domain. In order to assess the construct validity of specific measures, or sets of measures, it is useful to speculate about the relationship between each measure and the dimensions of the performance domain as depicted in Figure 10.1. In order to illustrate this process, four distinct

Figure 10.2
On-Site vs. Off-Site Behaviors

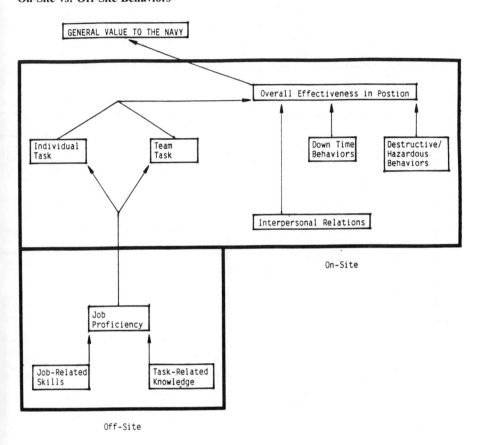

performance measurement technologies will be considered: (1) job
knowledge/skills tests, (2) hands-on testing, (3) simulations, and (4) rating
scales. There are several distinct applications of each technology; a description of some of these measures is listed below.

Job Knowledge/Skills Testing

1. *Paper and Pencil Tests.* Most likely to be used to measure task knowledge.
2. *Job Skills Tests.* Performance tests can be used to assess skill components of job
   performance. For example, a test of skill in using power tools could be developed
   by having examinees complete several standard operations using designated tools.
   The skills measured here are more basic and general than the specific tasks which
   are performed in doing individual jobs. For example, electronic troubleshooting
   might represent a skill, whereas repairing a sonar console might represent a task.

Figure 10.3
Task vs. Nontask Behaviors

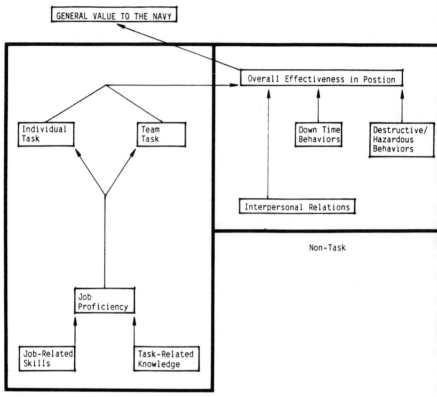

Hands-on Testing

3. *On-site hands-on testing.* Here, the subject performs one or more tasks in the normal work setting while being observed by an evaluator or evaluation team.
4. *Off-site hands-on testing.* The equivalent of work sample tests, in which the subject carries out tasks using normal equipment and techniques, but in a context that is different from the work site. Van tests provide an example of this category. This type of testing measures performance of whole tasks rather than of their basic skill or knowledge components.

Simulation

5. *High-fidelity simulations.* Similar to off-site hands-on testing, this technique attempts to duplicate the salient features of the work site and to obtain job proficiency measures under realistic conditions. Aircraft simulators provide an examle.
6. *Symbolic simulation.* Also related to hands-on testing, this technique involves the use of pictoral or video materials in depicting aspects of the task or job environment.

Figure 10.4
Cross-Classification of Behaviors

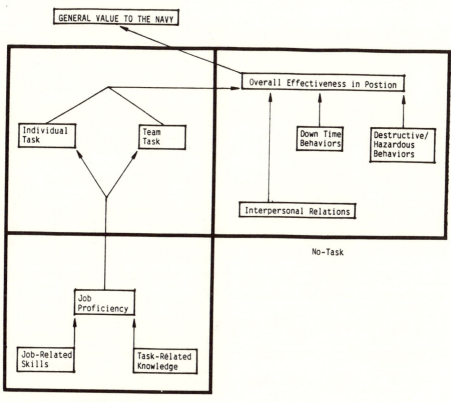

The subject's task is to indicate. using these symbolic materials, how he would carry out a task. As with hands-on tests described above, this technique measures performance in whole tasks rather than in basic task components.

Ratings

7. *Task Ratings.* Evaluations obtained from supervisors, peers, or the subject on how well or effectively a specific task is carried out. These judgments are distinct from the category that follows in that they are tied to individual tasks.

8. *Global Ratings.* Evolutions obtained from supervisors, peers, or the subject on general dimensions such as technical competence, safety etc. These represent evaluations of the person rather than of a specific task carried out by that person.

RELATING PERFORMANCE MEASURES TO DIMENSIONS

There is not, for the most part, a one-to-one correspondence between specific performance measures and work behaviors outlined in Figure 10.1.

We must therefore attempt to describe the relationships between each of the eight performance measures and the nine categories of work behavior identified in Figure 10.1. Initial estimates of these relationships are shown in in Table 10.1. These relationships can be characterized according to three dimensions: (1) Spurious—Casual, (2) Weak—Strong, and (3) Variable—Stable. The first dimension refers to the extent to which each of the criterion dimensions can be regarded as a direct cause of the behavior measured by the particular performance measure. The category "spurious" includes cases in which no relationship is expected; the category "causal" includes cases in which the behavior sampled by the performance measure is the same as the behavior that defines the criterion level. The second dimension refers to the expected value of the correlation between each performance measure and a hypothetical perfect measure of the work behavior. Note here that a strong relationship is not necessarily an indication of construct validity; substantial correlations may be the result of bias, shared method variance, etc. The third dimension refers to the extent to which this relationship is expected to be invariant across different contexts. Three-point scales are used for each dimension; a rating of 3, 3, 3 would indicate that the performance measure is directly caused by the criterion construct in question, and that this relationship is strong and invariant across the contexts which are likely to be encountered. The rationale for these proposed relations are presented below.

1. Paper and pencil tests: the benchmarks which define job knowledge. Knowledge is a component of job proficiency, but individual differenes in job knowledge are likely to reflect time in job and training level more than anything else, and thus will be unrelated to interpersonal and nontask behaviors, and only weakly related to task performance and overall effectiveness.

2. Job skills tests: the relationships proposed for job knowledge tests will also hold for this class of performance measures. One exception is in the relationship between skills and task performance. Since skills are likely to be more general and transportable than specific task knowledge, skills will be more strongly related to performance in a variety of post-training contexts than will measures of job knowledge.

3. On-site hands-on testing: the benchmark which defines job proficiency. Proficiency, in turn, is in part the outcome of training, job knowledge, and job skills. There in no reason to expect a consistent relationship between this performance measure and nontask behaviors such as accidents, absenteeism, or interpersonal conflict (the same can be said for the next three performance measures). The relationship between this measure and criteria at high levels in the framework is uncertain, since ability to accomplish tasks under controlled conditions does not always translate into effective job performance.

4. Off-site hands-on testing: since testing is removed from the usual context, this measure will not be as strongly related to job proficiency as will on-site measures. On the other hand, if contextual cues are removed, more generalized job skills and knowledge may play a greater role. For all other criterion levels, comments made with regard to on-site testing apply here also.

Table 10.1
Relationships between Performance Measures and Work Behaviors Described in Terms of Causality, Strength, and Invariance

RATING SCALES

Causality

1	2	3
Spurious		Causal

Strength

1	2	3
Weak		Strong

Stability

1	2	3
Variable		Stable

Performance Measures

	Paper/ Pencil	Job Skills	On-site Hands-on	Off-site Hands-on	Hi-Fi Simul	Symbol Simulat	Task Rat.	Global Rating
Overall Effectiveness	1,1,2	1,1,2	1,2,2	1,2,2	1,2,2	1,1,2	1,2,3	3,2,3
Task Performance	1,1,2	1,2,2	1,2,1	1,2,1	1,2,1	1,2,1,	3,3,2	2,3,2
Non-Task Behaviors	1,1,3	1,1,3	1,1,3	1,1,3	1,1,3	1,1,3	1,2,2	3,3,3
Interpersonal Relations	1,1,3	1,1,3	1,1,3	1,1,3	1,1,3	1,1,3	1,2,2	2,3,2
Job Proficiency	2,2,2	2,2,2	3,3,3	3,2,3	3,3,3	2,2,2	2,2,1	2,2,2
Job Skills/ Knowledge	3,3,3	3,3,3	3,2,2	3,3,2	3,3,2	3,3,3	2,2,2	2,2,2

5. High-fidelity simulations: although similar to on-site hands-on tests, these measures deprive the subject of specific contextual cues, such as the physical layout of work, or the placement of machines and structures, and thus may lead to greater emphasis on job-related skills and knowledge. However, if sufficient fidelity can be achieved, this technique would lead to less of an emphasis on these same factors than would off-site hands-on tests.

6. Symbolic simulations: since these are a step removed from acutal hands-on testing, these measures may relate more strongly to training, and possible to verbal and spatial abilitiy, and less strongly to job proficiency than any of the hands-on or high-fidelity simulation methods.

7. Task ratings: judgmental measures, even when tied to specific tasks, are likely to be biased in the direction of the rater's general evaluation of the ratee (Halo effect). This will vary by source (supervisor, peer, self) but is likely to lead to consistent positive correlations between ratings (both task and global) and assessments of the subject's overall effectiveness. This, in turn, will lead to a positive correlation between task ratings and assessments of the subject's overall value to the Navy. Assessments of training performance and job knowledge/skills are likely to bias task ratings, as are nontask behaviors. Finally, task ratings will be strongly related to task performance; this relationship will be stronger for ratings of individual performance than to ratings of performance as a team member, since many facets of team performance are beyond the control of the individual ratee.

8. Global ratings: will be strongly affected by both overall task performance and nontask behaviors. Training performance will bias global ratings, particularly if training scores are known to the rater. Finally, proficiency, skills, and knowledge are all indirectly related to global ratings. Positive correlations would be expected between global ratings.

Interpreting Table 10.1

Table 10.1 portrays a set of complex, multifacted relationships between actual performance measures and performance constructs (work behaviors). Two comments should be made regarding the interpretaion of these relationships. First, for different purposes one, two, or all three dimensions might be used to characterize these relationships. Different conclusions regarding the mapping of performance measures onto the criterion space could be reached, depending on which combination of dimensions was used. Second, this set of relationships exemplifies the bandwidth-fidelity trade-off. For example, all classes of work behaviors are thought to have a moderate to strong effect on global ratings. This indicates that global ratings could be used as a partial measure of any of the major classes of work behavior. On the other hand, it would be difficult to uniquely identify global ratings with any particular type or level of criterion. Hands-on tests, on the other hand, are uniquely identified with a job skills and proficiency, but may not provide valid measures of other behaviors. If performance is very narrowly defined, one would hope to attain an analog of simple structure in the set of relationships depicted in Table 10.1. That is, under a very tight definition of performance, it would be best to develop sets of performance measures which are identified uniquely with one performance construct. In this case, the tendency of more global measures to "load" on many work behavior dimensions would represent a form of criterion contamination. On the other hand, if performance is defined very broadly, spanning large segments of the domain of work behaviors, the tendency of several measures to "load" on only one or two work behavior dimensions might be regarded as criterion insufficiency.

Measurement Implications

Table 10.1 implies that different performance measures will be best suited for measuring quite different aspects of the performance domain. For example, both paper and pencil tests and performance tests of basic skills measure job knowledge/skills and provide information which is relevant in assessing proficiency, but are only tangentially useful to higher level criteria. Hands-on tests measure proficiency as well as knowledge and skills; off-site tasks may relate more strongly to general skills and less strongly to proficiency since they deprive the examinee of specific job-related cues provided by the work environment.

High-fidelity simulations are essentially similar to on-site hands-on tests. They may, however, place demands on some general skills, since they also deprive the examinee of specific contextual cues which may be used in carrying out the job. Symbolic simulations may provide higher levels of fidelity in some jobs more than in others, the ratings in Table 10.1 are based on the assumption that symbolic simulations involve significant abstraction, and thus may tap more general skills in the same way as paper and pencil tests, and off-site hands-on tests do.

Taken as a group, paper and pencil tests, job skills tests, hands-on tests, and simulations all measure the skills/knowledge/proficiency domain. Performance measures differ in the extent to which they place a premium on specific or general skills and knowledge, but do not differ in their essentials.

Task ratings are likely to be tied to task performance, although they will be influenced by rater biases and by the extent to which raters actually observe the tasks rated. Nontask factors will have their greatest impact when task outcomes are not highly visible (for example, some inspection tasks), and when raters have limited opportunnities to observe actual task performance. Under these circumstances, task ratings will reflect global evaluations of ratees. Global ratings, in turn, represent the only feasible measure of overall effectiveness. The problem with global ratings is that they are affected by task behaviors, nontask behaviors, and by a host of rater and ratee characteristics. Thus, it is difficult to know precisely what global ratings measure.

DERIVING PERFORMANCE DIMENSIONS

The framework shown in Figure 10.1 presents a rather general set of dimensions that define the construct of job performance for an extremely large and varied set of jobs—the enlisted jobs in the U.S. Navy. It is likely that different dimensions would emerge if the performance domain were defined for some different job or set of jobs. Indeed, it is likely that different dimensions would emerge for specific subsets of the job group covered in Figure 10.1 (example, shore versus aviation jobs on ships). It is useful to describe processes by which performance dimensions can be defined for any given job or set of jobs, so that framework similar to that shown in Figure 10.1 can be developed in other contexts.

In deriving performance dimensions, there are three specific dimensions that are likely to be relevant for *any* job or set of jobs. First is task accomplishment. The specific nature of this dimension will vary from job to job, and in some jobs it may be difficult to define the set of tasks that are associated with the job. Nevertheless, it is reasonable to assume that in any job there are several concrete tasks that the incumbent is expected to complete, and that the incumbent's success in completing these tasks will be a major determinant of job performance. Second, the category referred to in Figure 10.1 as "down-time" behaviors is likely to be relevant in almost every job. The importance of this category of behaviors will vary depending on the interdependence among work team members, the flexibility of work schedules, the predictability of job demands, and the mental and physical demands that are inherent in the job. In almost any case, however, the tendency of some workers to avoid the work setting or to come to work in some impaired state will affect their job performance. Third, the dimension labeled "interpersonal relations" is likely to be relevant in almost every work setting. Even the individual artist who works in his or her own home must eventually deal with customers, and the artist who is sufficiently abrasive may starve, regardless of talent. Most workers must maintain satisfactory interpersonal relations with coworkers, supervisors and subordinates; they may also have to deal with members of other organizational units, with customers, with the public, etc. Although smooth interpersonal relations do not guarantee good performance, the worker who is unable to get along with the boss, coworkers, or customers is not likely to be an effective performer.

Task accomplishment, down-time behaviors, and interpersonal relations can be regarded as essentially universal dimensions of job performance. The specific behaviors that are included in these categories, the identity of other relevant categories, and the relative emphasis placed on other performance dimensions will vary depending on the job(s) in question, and more important, on the level of analysis.

Level of Analysis

The framework illustrated in Figure 10.1 covered a large and diverse set of jobs. As a result, the dimensions used to define the performance domain were both broad and general. Application of a similar process to a single job, or a homogeneous set of jobs, might yield performance dimensions that were considerably more specific and refined.

It is not clear that there is any advantage to deciding what level of analysis is the "correct" level for defining the performance domain. Different levels of analysis may be useful for different purposes. Thus, it may be inappropriate to ask what are *the* dimensions of performance. It is clear, however, that the level of analysis will affect the sorts of goals that are defined as relevant, and therefore will affect the specification of the behaviors that are most relevant to those goals (that is, performance dimensions). First, there is evidence that the goals of an organization are generally different from (and more specific than) those than the organization's official or implicit mission statement (Rice, 1963). That is, organizational goals cannot be specified in any detail simply by knowing the general mission or the organization (for example, profit, service, regulation, protection). Second, the goals of units within organizations are not necessarily related to organizational goals (Hunt, 1976; Trist, Higgins, Murray, & Pollack, 1963). For example, the owners of a mine might set goals that maximize production, given a certain margin for safety, but work group goals may maximize safety, given a certain margin for performance. These two sets of goals will yield different ideas of what performance is, of whether workers are performing well or not, and of the extent, the nature, and the causes of individual differences in job performance.

Deriving Goals

Once the relevant goals of an organization or an organizational unit have been defined, the process of defining performance dimensions is relatively straightforward. The definition of these goals, however, could be a complex matter, especially since the professed goals of a work unit might be quite different than the goals actually pursued by that unit (that is, operating goals) (Perrow, 1961).

A sociotechnical systems approach might provide the best method for deriving performance goals and performance dimensions. This approach assumes that the nature and definition of a job, and of the work carried out by an incumbent, depends on both the social and the technological organization of work behavior (Emery & Trist, 1960; Herbst, 1974; Rice, 1958; Susman, 1976; Trist & Bamforth, 1951; Woodward, 1965, 1970). The sociotechnical analysis of work stresses the detailed observation of intact work units, with particular attention given to the interdependence between workers and work groups, the interplay between technology and the methods employed by individuals and work groups to accomplish major tasks, and to the effects of changes in either the social or the technical milieu on the work and on the worker. An example of the application of sociotechnical approach will be taken from the classic study of the interaction between social and technical factors in underground coal mining done by Trist et al. (1963).

The basic tasks carried out in a coal mine are to free the coal from the work face, to transport that coal out of the mine, and to erect and maintain the system of braces, timbers, and tunnels that allow workers to safely work underground. The traditional method of coal mining, referred to as the "single place" method, was one in which small groups worked on individual faces. In this method, groups were self-sufficent and self-regulating. That is, each group was responsible for carrying out all of the tasks involved in mining, and there was minimal contact between groups. This method called for multiple skills and multiple task roles; senior miners were not specialized, but rather were expected to perform a wide range of tasks, from extracting and shoveling coal to building and maintaining roof supports. The development of the mechanical scrapers and flexible conveyer belts led to the development of the "longwall" method, which entailed a radical change in the nature and organization of work.

The longwall method is one in which large groups of miners (for example, 40 or more) work a continuous face of 80 to 100 yards. This method involves the formal separation of the three major processes of preparation, getting, and advancing, and has led to substantial specialization in miners' jobs. In the longwall method, the production cycle runs 24 hours, with three separate shifts carrying out completely different tasks. In the first shift, the coal face is undercut to a depth of four to six feet, and explosives are used to free the coal. The second shift is respononsible for removing the coal and for building roof supports. The third shift is responsible for advancing the conveyors and cutters to the new face, and for preparing the tunnels and gateways for further cutting. At the end of 24 hours, the coal face is ready to be undercut, and the cycle begins again.

The change from the single place to the longwall method of mining had a substantial effect on the nature of the miners' work and on the definition of job performance. In the single place method, senior miners were responsible for deciding the pace of work, the strategy for mining a particular seam, and the allocation of tasks within the work group. All experienced miners were expected to carry out a wide range of tasks, depending on the current conditions of the face. Thus, for a single-place miner, a major determinant of job performance was the ability to analyze coal face conditions and determine and put into action the optimal strategy for safely removing coal. In addition, physical effort had a major impact on productivity. Trist et al. (1963) noted that work groups tended to be homogeneous in many respects, and that groups consisting of strong, physically fit miners produced (and were paid) twice or even three times as much as less fit groups. Finally, performance in a single place mine depended more on the worker's ability to carry out a variety of tasks than on his proficiency with any single task.

In a single place mine, the pace of work varied tremedously across work groups, but the overall flow of coal from the mine was fairly steady, and was not strongly affected by slowdowns, accidents, or other work stoppages in any work group. In contrast, a longwall mine features a high level of interdependence among shifts. If any one shift is unable to complete its task, the entire production cycle is interrupted, and the productivity of the mine is substantially effected. For example, if the third shift is unable to successfully move all of the equipment to the face, the first and second shifts will not be able to complete their tasks. Since workers on each shift are highly specialized, it might take an entire day before all of the third-shift tasks are completed and the production cycle is ready to resume.

The longwall method places a premium on dependability and on ability to adapt specialized functions to the changing conditions of an underground mine. In a longwall mine, the avoidance of down-time behaviors is a critical aspect of performance. In addition, the ability to carry out a specific task (for example, operate cutting machinery) under varying conditions is important. Futhermore, effective communications between workers who occupy different jobs, and between workers on different shifts, is critical in the longwall method. In a single place mine, more emphasis must be placed on maintaining interpersonal relations within a small group; in a longwall mine the group is larger, and the emphasis is on communicating information rather than on personal relations.

In a longwall mine, there is no immediate relationship between the level effort exerted by the individual miner and the productivity of the work group. Rather, the primary emphasis in the longwall mine is the avoidance of breakdowns, accidents, or work conditions that will prevent a shift from completing its appointed task. In a single place mine, the miner's job is to choose the task or method that is most appropriate for the situation at hand. In the longwall method, the worker's job is to guarantee that a single well-defined task is carried out in a particular span of time.

SUMMARY

The central thesis of this chapter is that the dimensions of job performance must be specified in order to make significant progress in the prediction, measurement, and understanding of job performance. Futhermore, it is argued that performance dimensions cannot be discovered through some statistical slight-of-hand, but rather must be defined. These dimensions are, in turn, an outgrowth of the goals that are pursued by the organization, the work group, the incumbent, etc. The dimensions of job performance are those sets of behaviors that are directly related to those goals.

The level of analysis will have a strong impact on the set of goals that direct the behavior of the organizational unit being analyzed. One implication of this is that there in no "correct" set of job performance dimensions. Depending on one's purpose and one's unit of analysis, very different dimensions might emerge. Using a set of performance dimensions derived for enlisted jobs in the Navy, it was shown how performance dimensions can be related to a variety of measurements technologies. The relations shown in Table 10.1 could be regarded as the first step in establishing the construct validity of specific performance measures.

In general, one must specify the set of relevant goals before it is possible to derive performance dimensions. There is a substantial body of research of social-technical systems that may provide methods of specifying these goals and of defining the dimensions of job performance.

NOTE

1. Not all methods of confirmatory factor analysis involve these assumptions (Bentler & Weeks, 1980; Joreskog, 1969; Long, 1983). However, any method of factor analysis imposes some statistical structure on the dimensions obtained.

REFERENCES

Astin, A. (1964). Criterion-centered research. *Educational and Psychological Measurement, 24,* 807-822.
Bass, B. (1982). Individual capability, team performance, and team productivity. In M. Dunnette and E. Fleishman (Eds.), *Human performance and productivity: Human capability assessment.* Hillsdale, NJ: Erlbaum.
Bentler, P. M., and Weeks, D. G. (1980). Linear structural equations with latent variables. *Psychometrika, 45,* 289-308.
Bialek, H., Zapf, D., and McGuire, W. (1977, June). *Personnel turbulence and time utilization in an infantry division* (Hum RRO FR-WD-CA 77-11). Alexandria, VA: Human Resources Research Organization.
Borman, W. C. (1977). Consistency of rating accuracy and rating errors in the judgment of human performance. *Organizational Behavior and Human Performance, 20,* 238-252.
Campbell, J. (1983). Some possible implications of "modeling" for the conceptualization of measurement. In F. Landy, S. Zedeck, and J. Cleveland (Eds.) *Performance measurement and theory.* Hillsdale, NJ: Erlbaum.
Campbell, J., Dunnette, M., Lawer, E., and Weick, K. (1970). *Managerial behavior, performance and effectiveness.* New York: McGraw-Hill.
Chapanis, A. (1976). Engineering psychology. In M. Dunnette (Ed.), *Handbook of industrial and organizational psychology.* Chicago: Rand McNally.
Christal, R. E. (1974). *The United States Air Force occupational research project* (AFHRL-TR-73-75). Lackland AFB, TX: USAF, AFHRL, Occupational Research Division.

Cooper, W. (1981). Ubiquitous halo: Sources, solutions, and a paradox. *Psychological Bulletin, 90,* 218-244.

Dunnette, M. (1963). A note on the criterion. *Journal of Applied Psychology, 47,* 251-254.

———— (1976). Aptitudes, abilities and skills. In M. Dunnette (Ed.), *Handbook of industrial and organizational psychology.* Chicago: Rand McNally.

Emery, F. E., and Trist, E. L. (1960). Socio-technical systems. In C. Churchman and M. Verhurst (Eds.), *Management science, models, and techniques.* London: Pergamon Press.

Fleishman, E., and Quaintance, M. (1984). *Taxonomies of human performance: The description of human tasks.* New York: Academic Press.

Guion, R. M. (1980). On tinitarian doctrines of validity. *Professional Psychology, 11,* 385-398.

Guzzo, R. A., Jette, R. D., and Katzell, R. A. (1985). The effects of psychologically-based intervention programs on worker productivity. *Personnel Psychology, 38,* 275-293.

Harvey, R. J. (1982). The future of partial correlation as means to reduce halo in performance ratings. *Journal of Applied Psychology, 67,* 171-176.

Hemphill, J. K. (1959). Job descriptions for executives. *Harvard Business Review, 37,* 55-67.

———— (1960). *Dimensions of executive positions* (Research monograph 98) Columbus: Bureau of Business Research, Ohio State University.

Herbst, P. (1974). *Socio-technical design: Strategies in multidisciplinary research.* London: Tavinstock Publications.

Hunt, R. G. (1976). On the work itself: Observations concerning relations between tasks and organizational processes. In E. Miller (Ed.), *Task and organization.* New York: Wiley.

Hunter, J. E., and Hunter, R. F. (1984). Validitiy and utility of alternative predictors of job performance. *Psychological Bulletin, 96,* 72-98.

Hunter, J. E., and Schmidt, F. L. (1982). Fitting people to jobs: The impact of personnel selection on national productivity. In M. Dunnette and E. Fleishman (Eds.), *Human performance and productivity: Human capability assessment.* Hillsdale, NJ: Erlbaum.

James, L. (1973). Criterion models and construct validity for criteria. *Psychological Bulletin, 80,* 75-83.

Joreskog, K. (1969). A general approach to confirmatory factor analysis. *Psychometrika, 34,* 183-202.

Kavanaugh, M. J. (1971). The content issue in performance appraisal: A review. *Personnel Psychology, 34,* 653-668.

Kavanaugh, M. J., MacKinney, A. C., and Wolins, L. (1971). Issues in managerial performance: Multitrait multimethod analyses of ratings. *Psychological Bulletin, 75,* 34-49.

Landy, F. J., and Farr, J. L. (1980). Performance rating. *Psychological Bulletin, 87,* 72-107.

———— (1983). *The measurement of work performance: Methods, theory and applications.* New York: Academic Press.

Landy, F. J., Vance, R. J., and Barnes-Farrell, J. L. (1982). Statistical control of halo: A response. *Journal of Applied Psychology, 67,* 177-180.

Landy, F. J., Vance, R. J., Barnes-Farrell, J. L., and Steelle, J. W. (1980). Statistical control of halo error in performance ratings. *Journal of Applied Psychology, 65,* 501-506.

Landy, F. J., Zedeck, S., and Cleveland, J. N. (1983). *Performance measurement and theory.* Hillsdale, NJ: Erlbaum.

Lawler, E. E. (1967). The multitrait-multi-rater approach to measuring managerial job performance. *Journal of Applied Psychology, 51,* 369-381.

Long, J. S. (1983). *Confirmatory factor analysis: A preface to LISREL.* Beverly Hills, CA: Sage.

McCormick, E. J. (1979). *Job analysis: Methods and application.* New York: Amacom.

McCormick, E. J., Jeanneret, P. R., and Mecham, R. C. (1972). A study of job characteristics and job dimensions as based on the Position Analysis Questionnaires (PAQ). *Journal of Applied Psychology, 56,* 347-368.

Mitchell, T. R., and Kalb, L. S. (1982). Effects of job experience on supervisors' attributions for a subordinate's poor performance. *Journal of Applied Psychology, 67,* 181-188.

Mitchell, T. R., and Wood, R. E. (1980). Supervisor's responses to subordinate poor performance: A test of an attributional model. *Organizational Behavior and Human Performance, 25,* 123-138.

Muckler, F. A. (1982). Evaluation productivity. In M. Dunnette and E. Fleishman (Eds.), *Human performance and productivity: Human capability assessment.* Hillsdale, NJ: Erlbaum.

Murphy, K. R. (1982). Difficulties in the statistical control of halo. *Journal of Applied Psychology, 67,* 161-164.

_____ (1985). *Dimensions of job performance.* Unpublished manuscript, Colorado State University.

_____ (1986). When your top choice turns you down: Effect of rejected offers on the utility of selection tests. *Psychological Bulletin, 99,* 133-138.

Murphy, K. R., and Balzer, W. K. (1981). *Rater errors and rating accuracy.* Presented at American Psychological Association Convention, Los Angeles.

Murphy, K. R., Garcia, M., Kerkar, M., Martin, C., and Balzer, W. (1982). Relationship between observational accuracy and accuracy in evaluating performance. *Journal of Applied Psychology, 67,* 320-325.

Nunnally, J. (1978). *Psychometric theory* (2nd ed). New York: McGraw-Hill.

Pearlman, K., Schmidt, F. L., and Hunter, J. E. (1980). Validity generalization results for tests used to predict job proficiency and training success in clerical occupations. *Journal of Applied Psychology, 65,* 373-406.

Perrow, C. (1961). An analysis of goals in complex organizations. *American Psychological Review, 26,* 854-866.

Pickle, H., and Friedlander, F. (1967). Seven societal criteria of oragnizational success. *Personnel Psychology, 20,* 165-178.

Rice, A. K. (1958). *Productivity and social organization.* London: Tavinstock Publications.

_____ (1963). *The enterprise and its environment: A systems theory of management.* London: Tavinstock Publications.

Richards, J., Taylor, C., Price, P., and Jacobsen, T. (1965). An investigation of the criterion problem for one group of medical specialists. *Journal of Applied Psychology, 49,* 79-90.

Ronan, W. (1963). A factor analysis of eleven job performance measures. *Personnel Psychology, 16,* 255-267.

Ronan, W. W., and Prien, E. P. (1966). *Toward a criterion theory: A review and analysis of research and opinion.* Greensboro, NC: The Richardson Foundation.

Rush, C. (1953). A factorial study of sales criteria. *Personnel Psychology, 6,* 9-24.

Saal, F., Downey, R., and Lahey, M. (1980). Rating the ratings: Assessing the psychometric quality of rating data. *Psychological Bulletin, 88,* 413-428.

Salvendy, G., and Seymor, W. (1973). *Prediction and development of industrial work performance.* New York: John Wiley & Sons.

Schmidt, F. L., and Hunter, J. E. (1977). Development of a general solution to the problem of validity generalization. *Journal of Applied Psychology, 62,* 529-540.

———. (1981). Employment testing: Old theories and new research findings. *American Psychologist, 36,* 1128-1137.

Schmidt, F. L., Hunter, J. E., McKenzie, R. C., and Muldrow, T. (1979). The impact of valid selection procedures on workforce productivity. *Journal of Applied Psychology, 64,* 609-626.

Schmidt, F. L., and Kaplan, L. D. (1971). Composite versus multiple criteria: A review and resolution of the controversies. *Personnel Psychology, 24,* 419-434.

Seashore, S. (1975). Defining and measuring the quality of working life. In L. Davis and A. Cherns (Eds.), *The quality of working life* (Vol. 1). New York: The Free Press.

Smith, P. C. (1976). Behaviors, results, and organizational effectiveness: The problem of criteria. In M. Dunnette (Ed.), *Handbook of industrial and organinzational psychology.* Chicago: Rand McNally.

Susman, G. I. (1976). *Autonomy at work: A sociotechnical analysis of participative management.* New York: Praeger.

Sutermeister, R. A. (1976). *People and productivity.* New York: McGraw-Hill.

Tornow, W. W., and Pinto, P. R. (1976). The development of a managerial job taxonomy: A system for describing, classifying, and evaluating, executive positions. *Journal of Applied Psychology, 61,* 410-418.

Trist, E. L., and Bamforth, K. W. (1951). Some social and psychological conseqences of the longwall method of goal-setting. *Human Relations, 4,* 3-38.

Trist, E. L., Higgins, G. W., Murray, H., and Pollock, A. B. (1963). *Organizational choice.* London: Tavinstock Publications.

Turner, W. (1960). Dimensions of foreman performance: A factor analysis of criterion measures. *Journal of Applied Psychology, 44,* 216-223.

Wallace, S. R. (1965). Criteria for what? *American Psychologist, 20,* 411-418.

Woodward, J. (1965). *Industrial organizations: Theory and practice.* London: Oxford University Press.

———. (1970). *Industrial organizations: Behavior and control.* London: Oxford University Press.

Index

ABCA questionnaire, 206–7
ability testing: and brain activity, 104–7; definition of, 100. *See also* cognitive skills modeling; individual differences; testing
accessibility, as measurement of knowledge, 161
achievement testing, 132–43; analytical and logical reasoning, 140–43; critical thinking, 133–36; definition of achievement, 100; objectives of, 133; problem-solving skills, 133–36; writing skills, 136–40
ACT, 108, 111
Adelson, B., 161
Advanced Placement Test, 137
AFQT (Armed Forces Qualification Test), 17, 28–31, 99, 108–12; norming and reference population for ASVAB and, 28–29; purpose and uses of, 24–25
AGCT (Army General Classification Test), 28–29
agreeableness, VZM measurement of, 205–6
Air Defense Radar Simulation Task (AIRDEF), 117–19
Air Force Office of Scientific Research (AFOSR), 147

Air Force's Learning Abilities Measurement Program (LAMP). *See* LAMP
Akkerman, A. E., 206
Alderton, D., 175, 179
amplitude measurements, of brain activity, 107–9
analytical reasoning, testing of, 140–43
Anastasi, A., 209
Anderson, J. R., 155, 158–60, 164–65, 168
Anderson, R. C., 39–40, 46, 160
Anderson, S. B., 11
Angoff, W. H., 81
antireductionist position: ROMAT and, 210–13; RSIC test and, 212–14
Appel, V., 11
Applebee, A. N., 138
aptitude testing, 1–13, 147–48; cognitive theory and, 147–48; criterion development, 9–10; criterion-related validity, 12; dimension measures, 6–9; evaluation of approaches, 10–13; implementation, 11–12; measures taken from, 5–9; method of administration, 2–5; new approaches, 1–13; practical utility, 11–12; psychodiagnositic utility, 11; replicability, 12; scoring, 11–12; standardizability, 12; theoretical motivation, 10–11